RF CIRCUIT DESIGN TECHNIQUES for MF-UHF APPLICATIONS

T0225390

RF
CIRCUIT DESIGN
TECHNIQUES
for MF-UHF
APPLICATIONS

Abdullah Eroglu

CRC Press
Taylor & Francis Group
Boca Raton London New York

CRC Press is an imprint of the
Taylor & Francis Group, an **informa** business

CRC Press
Taylor & Francis Group
6000 Broken Sound Parkway NW, Suite 300
Boca Raton, FL 33487-2742

First issued in paperback 2017

© 2013 by Taylor & Francis Group, LLC
CRC Press is an imprint of Taylor & Francis Group, an Informa business

No claim to original U.S. Government works
Version Date: 20130208

ISBN 13: 978-1-138-07760-7 (pbk)
ISBN 13: 978-1-4398-6165-3 (hbk)

Library of Congress Cataloging-in-Publication Data

Eroglu, Abdullah.
    RF circuit design techniques for MF-UHF applications / Abdullah Eroglu.
    pages cm
    Includes bibliographical references and index.
    ISBN 978-1-4398-6165-3 (hardback)
    1. Radio circuits. I. Title.

TK6560.E725 2013
621.3841'2--dc23   2012050935

Visit the Taylor & Francis Web site at
http://www.taylorandfrancis.com

and the CRC Press Web site at
http://www.crcpress.com

Dedicated to my mother, Gulser, and father, Ahmet Eroglu

Contents

Preface

RF/microwave engineering plays a vital role in our daily lives. From garage openers to door access, and cell phones to GPS devices, we use and experience devices that are engineered with RF technology everyday. RF technology also engineers several advanced devices and components that are used in military and industrial applications including semiconductor, biomedical, RFID, and radar.

Several of the important RF applications take place within the frequency range known as the medium frequency (MF) to ultrahigh frequency (UHF) range. Industrial, scientific, and medical applications, including medical resonance imaging, semiconductor processing, and RFID, are some of the critical applications within the MF-UHF range that require analytical and experimental RF techniques to understand the technology. Furthermore, the designer needs these techniques to design components, devices, and integrate them when needed with high efficiency, minimal loss, and required power. The task for RF engineers becomes more challenging with the advancement of technology. It is now a requirement for engineers to not only design high-performance devices, but it is also important to design them in a cost-effective way. That is why computer-aided design (CAD) tools are so important in the design and implementation of components and devices since they reduce the associated cost and improve the accuracy by optimization.

This book is intended to provide engineers and students the required skill set to design, simulate, and implement RF/microwave components and devices using theory and practice for applications within the MF-UHF range. Every section in the book first presents the required theory and then the verification process with CAD tools. This book has several real life implementation examples that complement theory and simulation and that make it unique. Design tables, curves, and charts based on the presented theory are uniquely prepared for engineers and students to present an efficient design process. CAD tools such as MATLAB®, Mathcad, HFSS, Ansoft Designer, Sonnet, and Pspice are used in several examples presented in this book.

The scope of each chapter in the book can be summarized as follows. Chapter 1 presents network parameters that are frequently used in the analysis of RF components and devices. Network parameters including Z, Y, h, $ABCD$, and scattering parameters are discussed in detail in this chapter. Inductor design techniques, including microstrip and composite inductors, are presented in Chapter 2. Chapter 3 presents transformer design and implementation methods. The design and implementation of conventional transformers, autotransformers, and transmission line transformers are discussed in detail with examples. Combiner, divider, and phase inverters using lumped elements and distributed elements are presented in Chapter 4. Unique directional coupler design methods using lumped elements and transmission lines are discussed in detail with examples in Chapter 5. Filter design with several examples is presented in Chapter 6. Chapter 7 describes the RFID system design and details microstrip-type RFID antennas and their simulation. EBG-type RFID microstrip antennas are also discussed in this chapter.

MATLAB® is a registered trademark of The MathWorks, Inc. For product information, please contact:

The MathWorks, Inc.
3 Apple Hill Drive
Natick, MA 01760-2098 USA
Tel: 508-647-7000
Fax: 508-647-7001
E-mail: info@mathworks.com
Web: www.mathworks.com

Acknowledgments

I would like to thank my wife, G. Dilek, for her support and editorial corrections in the book. I would also like to acknowledge my students at IPFW for their help in the preparation of some of the examples. Special thanks goes to my editor, Nora Konopka, for her patience and support during the course of the preparation and publication of this book.

It is impossible not to recognize the dedication shown by my wife and children. They always supported and gave me the strength when I needed to complete every task I started.

1 Network Parameters in RF Circuit Design

1.1 INTRODUCTION

Network parameters allow engineers to determine the overall circuit performance without knowing the internal structure. They carry great importance in the analysis and design of devices and components. Network parameters provide mathematical tools for designers to model and characterize devices by establishing relations between voltages and currents [1–5]. It is possible to theoretically calculate loss, power delivered, reflection coefficient, voltage and current gains, and several other critical parameters with the use of network analysis techniques. Hence, it is necessary to understand and utilize network parameters in radio frequency (RF)/microwave device and component design to have better performance.

1.2 NETWORK PARAMETERS

The analysis of network parameters can be explained using two-port networks. The two-port network shown in Figure 1.1 is described by a set of four independent parameters, which can be related to voltage and current at any port of the network. As a result, two-port network can be treated as a black box modeled by the relationships between the four variables. There exist six different ways to describe the relationships between these variables depending on which two of the four variables are given, while the other two can always be derived. All voltages and currents are complex variables and are represented by phasors containing both magnitude and phase. Two-port networks are characterized using two-port network parameters such as Z-impedance, Y-admittance, h-hybrid, and $ABCD$. They are usually expressed in matrix notation and they establish relations between the following parameters: input

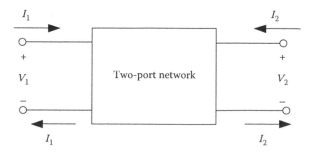

FIGURE 1.1 Two-port network representation.

voltage V_1, output voltage V_2, input current I_1, and output current I_2. High-frequency networks are characterized by S parameters.

1.2.1 Z-IMPEDANCE PARAMETERS

The voltages are represented in terms of currents through Z parameters as follows:

$$V_1 = Z_{11}I_1 + Z_{12}I_2 \tag{1.1}$$

$$V_2 = Z_{21}I_1 + Z_{22}I_2 \tag{1.2}$$

In matrix form, Equations 1.1 and 1.2 can be combined and written as

$$\begin{bmatrix} V_1 \\ V_2 \end{bmatrix} = \begin{bmatrix} Z_{11} & Z_{12} \\ Z_{21} & Z_{22} \end{bmatrix} \begin{bmatrix} I_1 \\ I_2 \end{bmatrix} \tag{1.3}$$

The Z parameters for a two-port network are defined as

$$Z_{11} = \left. \frac{V_1}{I_1} \right|_{I_2=0} \quad Z_{12} = \left. \frac{V_1}{I_2} \right|_{I_1=0}$$
$$Z_{21} = \left. \frac{V_2}{I_1} \right|_{I_2=0} \quad Z_{22} = \left. \frac{V_2}{I_2} \right|_{I_1=0} \tag{1.4}$$

The formulation in Equation 1.4 can be generalized for N-port network as

$$Z_{nm} = \left. \frac{V_n}{I_m} \right|_{I_k=0(k \neq m)} \tag{1.5}$$

Z_{nm} is the input impedance seen looking into port n, when all other ports are open circuited. In other words, Z_{nm} is the transfer impedance between ports n and m when all other ports are open. It can be shown that for reciprocal networks

$$Z_{nm} = Z_{mn} \tag{1.6}$$

1.2.2 Y-ADMITTANCE PARAMETERS

The currents are related to voltages through Y parameters as follows:

$$I_1 = Y_{11}V_1 + Y_{12}V_2 \tag{1.7}$$

$$I_2 = Y_{21}V_1 + Y_{22}V_2 \tag{1.8}$$

In matrix form, Equations 1.7 and 1.8 can be written as

$$\begin{bmatrix} I_1 \\ I_2 \end{bmatrix} = \begin{bmatrix} Y_{11} & Y_{12} \\ Y_{21} & Y_{22} \end{bmatrix} \begin{bmatrix} V_1 \\ V_2 \end{bmatrix} \tag{1.9}$$

The Y parameters in Equation 1.9 can be defined as

$$Y_{11} = \frac{I_1}{V_1}\Big|_{V_2=0} \qquad Y_{12} = \frac{I_1}{V_2}\Big|_{V_1=0}$$

$$Y_{21} = \frac{I_2}{V_1}\Big|_{V_2=0} \qquad Y_{22} = \frac{I_2}{V_2}\Big|_{V_1=0} \tag{1.10}$$

The formulation in Equation 1.10 can be generalized for N-port network as

$$Y_{nm} = \frac{I_n}{V_m}\Big|_{V_k=0(k \neq m)} \tag{1.11}$$

Y_{nm} is the input admittance seen looking into port n, when all other ports are short circuited. In other words, Y_{nm} is the transfer admittance between ports n and m when all other ports are short. It can be shown that for reciprocal networks

$$Y_{nm} = Y_{mn} \tag{1.12}$$

In addition, it can be further proven that the impedance and admittance matrices are related through

$$[Z] = [Y]^{-1} \tag{1.13}$$

or

$$[Y] = [Z]^{-1} \tag{1.14}$$

1.2.3 ABCD PARAMETERS

ABCD parameters relate voltages to currents in the following form for two-port networks:

$$V_1 = AV_1 - BI_2 \tag{1.15}$$

$$I_1 = CV_1 - DI_2 \tag{1.16}$$

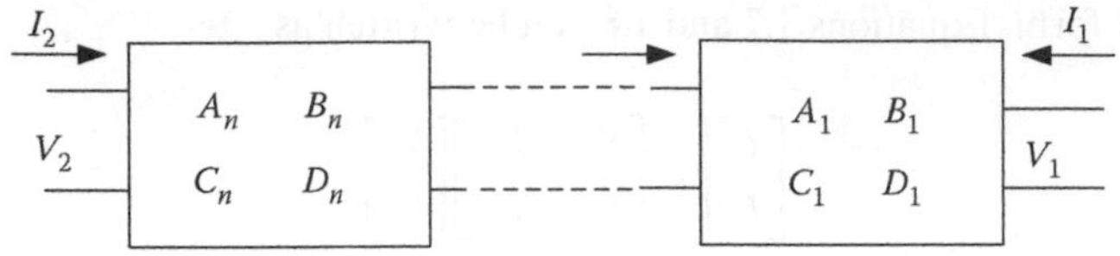

FIGURE 1.2 *ABCD* parameter of cascaded networks.

which can be put in the matrix form as

$$
\begin{bmatrix} V_1 \\ I_1 \end{bmatrix} = \begin{bmatrix} A & B \\ C & D \end{bmatrix} \begin{bmatrix} V_1 \\ -I_2 \end{bmatrix}
\tag{1.17}
$$

ABCD parameters in Equation 1.17 are defined by

$$
A = \frac{V_1}{V_2}\bigg|_{I_2=0} \qquad B = \frac{V_1}{-I_2}\bigg|_{V_2=0}
$$

$$
C = \frac{I_1}{V_2}\bigg|_{I_2=0} \qquad D = \frac{I_1}{-I_2}\bigg|_{V_2=0}
\tag{1.18}
$$

When the network is reciprocal, it can be shown that

$$
AD - BC = 1
\tag{1.19}
$$

$A = D$ for symmetrical network. *ABCD* parameters are useful in finding voltage or current gain of the component or the overall gain of a network. One of the greatest advantages of *ABCD* parameters is their use when networks or components are cascaded. When this condition exists, the overall *ABCD* parameter of the network simply becomes the matrix product of the individual network or components as given by Equation 1.20. This can be generalized for *N*-port network shown in Figure 1.2 as

$$
\begin{Bmatrix} v_1 \\ i_1 \end{Bmatrix} = \left(\begin{bmatrix} A_1 & B_1 \\ C_1 & D_1 \end{bmatrix} \cdots \begin{bmatrix} A_n & B_n \\ C_n & D_n \end{bmatrix} \right) \begin{Bmatrix} v_2 \\ -i_2 \end{Bmatrix}
\tag{1.20}
$$

1.2.4 *H*-HYBRID PARAMETERS

Hybrid parameters relate voltages and currents in a two-port network as

$$
V_1 = h_{11}I_1 + h_{12}V_2
\tag{1.21}
$$

$$
I_2 = h_{21}I_1 + h_{22}V_2
\tag{1.22}
$$

Equations 1.21 and 1.22 can be put in a matrix form as

$$\begin{bmatrix} V_1 \\ I_2 \end{bmatrix} = \begin{bmatrix} h_{11} & h_{12} \\ h_{21} & h_{22} \end{bmatrix} \begin{bmatrix} I_1 \\ V_2 \end{bmatrix}$$

(1.23)

The hybrid parameters in Equation 1.23 can be found from

$$h_{11} = \left. \frac{V_1}{I_1} \right|_{V_2=0} \qquad h_{12} = \left. \frac{V_1}{V_2} \right|_{I_1=0}$$

$$h_{21} = \left. \frac{I_2}{I_1} \right|_{V_2=0} \qquad h_{22} = \left. \frac{I_2}{V_2} \right|_{I_1=0}$$

(1.24)

Hybrid parameters are preferred for components such as transistors and transformers since they can be measured with ease in practice.

Example

Find the (a) impedance, (b) admittance, (c) *ABCD*, and (d) hybrid parameters of the T-network given in Figure 1.3.

SOLUTION

a. *Z* parameters are found with the application of Equation 1.4 by opening all other ports except the measurement port. This leads to

$$Z_{11} = \left. \frac{V_1}{I_1} \right|_{I_2=0} = Z_A + Z_C, \quad Z_{21} = \left. \frac{V_2}{I_1} \right|_{I_2=0} = Z_C$$

$$Z_{12} = \left. \frac{V_1}{I_2} \right|_{I_1=0} = \frac{V_2}{I_2} \frac{Z_C}{Z_B + Z_C} = (Z_B + Z_C)\frac{Z_C}{Z_B + Z_C} = Z_C, Z_{22} = \left. \frac{V_2}{I_2} \right|_{I_1=0} = Z_B + Z_C$$

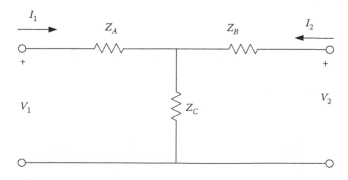

FIGURE 1.3 T-network configuration.

The Z matrix is then constructed as

$$
Z = \begin{bmatrix} Z_A + Z_C & (Z_B + Z_C)\dfrac{Z_C}{Z_B + Z_C} = Z_C \\[2ex] (Z_B + Z_C)\dfrac{Z_C}{Z_B + Z_C} = Z_C & Z_B + Z_C \end{bmatrix}
$$

b. Y parameters are found from Equation 1.10 by shorting all other ports except the measurement port. Y_{11} and Y_{21} are found when port 2 is shorted as

$$
Y_{11} = \left.\frac{I_1}{V_1}\right|_{V_2=0} \rightarrow I_1 = \frac{V_1}{Z_A + \left(Z_B \ // Z_C\right)} = V_1\left(\frac{Z_B + Z_C}{Z_A Z_B + Z_A Z_C + Z_B Z_C}\right)
$$

$$
Y_{11} = \left(\frac{Z_B + Z_C}{Z_A Z_B + Z_A Z_C + Z_B Z_C}\right)
$$

$$
Y_{21} = \left.\frac{I_2}{V_1}\right|_{V_2=0} \rightarrow I_2 = \frac{-V_1}{\left(Z_A + \left(Z_B \ // Z_C\right)\right)\left(Z_C + Z_B\right)} \rightarrow Y_{21} = \left(\frac{-Z_C}{Z_A Z_B + Z_A Z_C + Z_B Z_C}\right)
$$

Similarly, Y_{12} and Y_{22} are found when port 1 is shorted as

$$
Y_{12} = \left.\frac{I_1}{V_2}\right|_{V_1=0} \rightarrow I_1 = \frac{-V_2}{(Z_B + (Z_A \ // Z_C))(Z_A + Z_C)} \rightarrow Y_{12} = \left(\frac{-Z_C}{Z_A Z_B + Z_A Z_C + Z_B Z_C}\right)
$$

$$
Y_{22} = \left.\frac{I_2}{V_2}\right|_{V_1=0} \rightarrow I_2 = \frac{V_2}{Z_B + (Z_A \ // Z_C)} = V_1\left(\frac{Z_A + Z_C}{Z_A Z_B + Z_A Z_C + Z_B Z_C}\right) \rightarrow
$$

$$
Y_{22} = \left(\frac{Z_A + Z_C}{Z_A Z_B + Z_A Z_C + Z_B Z_C}\right)
$$

Y parameters can also be found by just inverting the Z matrix given by Equation 1.14 as

$$
[Y] = [Z]^{-1} = \frac{1}{(Z_A Z_B + Z_A Z_C + Z_B Z_C)}\begin{bmatrix} Z_B + Z_C & -Z_C \\ -Z_C & Z_A + Z_C \end{bmatrix}
$$

So, the Y matrix for T-network is then

$$
Y = \begin{bmatrix} \left(\dfrac{Z_B + Z_C}{Z_A Z_B + Z_A Z_C + Z_B Z_C}\right) & \left(\dfrac{-Z_C}{Z_A Z_B + Z_A Z_C + Z_B Z_C}\right) \\[3ex] \left(\dfrac{-Z_C}{Z_A Z_B + Z_A Z_C + Z_B Z_C}\right) & \left(\dfrac{Z_A + Z_C}{Z_A Z_B + Z_A Z_C + Z_B Z_C}\right) \end{bmatrix}
$$

As seen from the results of part (a) and (b), the network is reciprocal since

$$Z_{12} = Z_{21} \quad \text{and} \quad Y_{12} = Y_{21}$$

c. Hybrid parameters are found using Equations 1.24. Parameters h_{11} and h_{21} are obtained when port 2 is shorted as

$$h_{11} = \frac{V_1}{I_1}\bigg|_{V_2=0} \rightarrow V_1 = I_1(Z_A + (Z_B \,//\,Z_C)) = I_1\left(\frac{Z_AZ_B + Z_AZ_C + Z_BZ_C}{Z_B + Z_C}\right) \rightarrow$$

$$h_{11} = \left(\frac{Z_AZ_B + Z_AZ_C + Z_BZ_C}{Z_B + Z_C}\right)$$

and

$$h_{21} = \frac{I_2}{I_1}\bigg|_{V_2=0} \rightarrow I_2 = -I_1\left(\frac{Z_C}{Z_B + Z_C}\right) \rightarrow h_{21} = -\left(\frac{Z_C}{Z_B + Z_C}\right)$$

Parameters h_{12} and h_{22} are obtained when port 1 is open circuited as

$$h_{12} = \frac{V_1}{V_1}\bigg|_{I_1=0} \rightarrow V_1 = V_2\left(\frac{Z_C}{Z_B + Z_C}\right) \rightarrow h_{12} = \left(\frac{Z_C}{Z_B + Z_C}\right)$$

and

$$h_{22} = \frac{I_2}{V_2}\bigg|_{I_1=0} \rightarrow I_2 = V_2\left(\frac{1}{Z_B + Z_C}\right) \rightarrow h_{22} = \left(\frac{1}{Z_B + Z_C}\right)$$

The hybrid matrix for T-network can now be constructed as

$$h = \begin{bmatrix} \left(\dfrac{Z_AZ_B + Z_AZ_C + Z_BZ_C}{Z_B + Z_C}\right) & \left(\dfrac{Z_C}{Z_B + Z_C}\right) \\[2ex] -\left(\dfrac{Z_C}{Z_B + Z_C}\right) & \left(\dfrac{1}{Z_B + Z_C}\right) \end{bmatrix}$$

d. *ABCD* parameters are found using Equations 1.18. Parameters *A* and *C* are determined when port 2 is open circuited as

$$A = \frac{V_1}{V_2}\bigg|_{I_2=0} \rightarrow V_2 = \frac{Z_C}{Z_C + Z_A}V_1 \rightarrow A = \left(\frac{Z_C + Z_A}{Z_C}\right)$$

and

$$C = \frac{I_1}{V_2}\bigg|_{I_2=0} \rightarrow I_1 = V_2\left(\frac{1}{Z_C}\right) \rightarrow C = \left(\frac{1}{Z_C}\right)$$

Parameters B and D are determined when port 2 is short circuited as

$$B = \frac{V_1}{-I_2}\bigg|_{V_2=0} \rightarrow I_2 = \frac{-V_1}{Z_A + (Z_B \; // \; Z_C)} \frac{Z_C}{(Z_B + Z_C)} \rightarrow B = \left(\frac{Z_A Z_B + Z_A Z_C + Z_B Z_C}{Z_C}\right)$$

and

$$D = \frac{-I_1}{I_2}\bigg|_{V_2=0} \rightarrow I_2 = -I_1\left(\frac{Z_C}{Z_B + Z_C}\right) \rightarrow D = \left(\frac{Z_B + Z_C}{Z_C}\right)$$

So, the *ABCD* matrix is found as

$$ABCD = \begin{bmatrix} \left(\dfrac{Z_C + Z_A}{Z_C}\right) & \left(\dfrac{Z_A Z_B + Z_A Z_C + Z_B Z_C}{Z_C}\right) \\ \left(\dfrac{1}{Z_C}\right) & \left(\dfrac{Z_B + Z_C}{Z_C}\right) \end{bmatrix}$$

It can be now proven from the results obtained that *Z*, *Y*, *h*, and *ABCD* parameters are related using the relations given in Table 1.1.

TABLE 1.1

Network Parameter Conversion Table

	[Z]	[Y]	[ABCD]	[h]
[Z]	$\begin{bmatrix} z_{11} & z_{12} \\ z_{21} & z_{22} \end{bmatrix}$	$\begin{bmatrix} \dfrac{y_{22}}{\Delta_Y} & \dfrac{-y_{12}}{\Delta_Y} \\ \dfrac{-y_{21}}{\Delta_Y} & \dfrac{y_{11}}{\Delta_Y} \end{bmatrix}$	$\begin{bmatrix} \dfrac{A}{C} & \dfrac{\Delta_T}{C} \\ \dfrac{1}{C} & \dfrac{D}{C} \end{bmatrix}$	$\begin{bmatrix} \dfrac{\Delta_H}{h_{22}} & \dfrac{h_{12}}{h_{22}} \\ \dfrac{-h_{21}}{h_{22}} & \dfrac{1}{h_{22}} \end{bmatrix}$
[Y]	$\begin{bmatrix} \dfrac{z_{22}}{\Delta_Z} & \dfrac{-z_{12}}{\Delta_Z} \\ \dfrac{-z_{21}}{\Delta_Z} & \dfrac{z_{11}}{\Delta_Z} \end{bmatrix}$	$\begin{bmatrix} y_{11} & y_{12} \\ y_{21} & y_{22} \end{bmatrix}$	$\begin{bmatrix} \dfrac{D}{B} & \dfrac{-\Delta_T}{B} \\ \dfrac{1}{B} & \dfrac{A}{B} \end{bmatrix}$	$\begin{bmatrix} \dfrac{1}{h_{11}} & \dfrac{-h_{12}}{h_{11}} \\ \dfrac{h_{21}}{h_{11}} & \dfrac{\Delta_H}{h_{11}} \end{bmatrix}$
[ABCD]	$\begin{bmatrix} \dfrac{z_{11}}{z_{21}} & \dfrac{\Delta_Z}{z_{21}} \\ \dfrac{1}{z_{21}} & \dfrac{z_{22}}{z_{21}} \end{bmatrix}$	$\begin{bmatrix} \dfrac{-y_{22}}{y_{21}} & \dfrac{-1}{y_{21}} \\ \dfrac{-\Delta_Y}{y_{21}} & \dfrac{-y_{11}}{y_{21}} \end{bmatrix}$	$\begin{bmatrix} A & B \\ C & D \end{bmatrix}$	$\begin{bmatrix} \dfrac{-\Delta_H}{h_{21}} & \dfrac{-h_{11}}{h_{21}} \\ \dfrac{-h_{22}}{h_{21}} & \dfrac{-1}{h_{21}} \end{bmatrix}$
[h]	$\begin{bmatrix} \dfrac{\Delta_Z}{z_{22}} & \dfrac{z_{12}}{z_{22}} \\ \dfrac{-z_{21}}{z_{22}} & \dfrac{1}{z_{22}} \end{bmatrix}$	$\begin{bmatrix} \dfrac{1}{y_{11}} & \dfrac{-y_{12}}{y_{11}} \\ \dfrac{y_{21}}{y_{11}} & \dfrac{\Delta_Y}{y_{11}} \end{bmatrix}$	$\begin{bmatrix} \dfrac{B}{D} & \dfrac{\Delta_T}{D} \\ \dfrac{1}{D} & \dfrac{C}{D} \end{bmatrix}$	$\begin{bmatrix} h_{11} & h_{12} \\ h_{21} & h_{22} \end{bmatrix}$

1.3 NETWORK CONNECTIONS

Networks and components in engineering applications can be connected in different ways to perform certain tasks. The commonly used network connection methods are series, parallel, and cascade connections. The series connection of two networks is shown in Figure 1.4a. Since the networks are connected in series, currents are the same and voltages are added across the ports of the network to find the overall voltage at the ports of the combined network. This can be represented by impedance matrices as

$$[Z] = [Z^x] + [Z^y] = \begin{bmatrix} Z_{11}^x & Z_{12}^x \\ Z_{21}^x & Z_{22}^x \end{bmatrix} + \begin{bmatrix} Z_{11}^y & Z_{12}^y \\ Z_{21}^y & Z_{22}^y \end{bmatrix} = \begin{bmatrix} Z_{11}^x + Z_{11}^y & Z_{12}^x + Z_{12}^y \\ Z_{21}^x + Z_{21}^y & Z_{22}^x + Z_{22}^y \end{bmatrix} \quad (1.25)$$

So

$$\begin{bmatrix} V_1 \\ V_2 \end{bmatrix} = \begin{bmatrix} Z_{11}^x + Z_{11}^y & Z_{12}^x + Z_{12}^y \\ Z_{21}^x + Z_{21}^y & Z_{22}^x + Z_{22}^y \end{bmatrix} \begin{bmatrix} I_1 \\ I_2 \end{bmatrix} \quad (1.26)$$

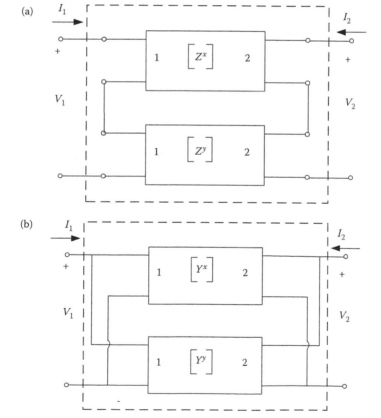

FIGURE 1.4 (a) Series connection of two-port networks. (b) Parallel connection of two-port networks.

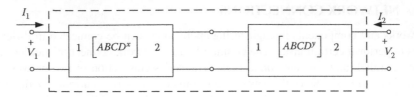

FIGURE 1.5 Cascade connection of two-port networks.

The parallel connection of two-port networks is illustrated in Figure 1.4b. In parallel-connected networks, voltages are the same across ports, and currents are added to find the overall current flowing at the ports of the combined network. This can be represented by Y matrices as

$$[Y] = [Y^x] + [Y^y] = \begin{bmatrix} Y_{11}^x & Y_{12}^x \\ Y_{21}^x & Y_{22}^x \end{bmatrix} + \begin{bmatrix} Y_{11}^y & Y_{12}^y \\ Z_{21}^y & Y_{22}^y \end{bmatrix} = \begin{bmatrix} Y_{11}^x + Y_{11}^y & Y_{12}^x + Y_{12}^y \\ Y_{21}^x + Y_{21}^y & Y_{22}^x + Y_{22}^y \end{bmatrix} \tag{1.27}$$

As a result

$$\begin{bmatrix} I_1 \\ I_2 \end{bmatrix} = \begin{bmatrix} Y_{11}^x + Y_{11}^y & Y_{12}^x + Y_{12}^y \\ Y_{21}^x + Y_{21}^y & Y_{22}^x + Y_{22}^y \end{bmatrix} \begin{bmatrix} V_1 \\ V_2 \end{bmatrix} \tag{1.28}$$

The cascade connection of two-port networks is shown in Figure 1.5. In the cascade connection, the magnitude of the current flowing at the output of the first network is equal to the current at the input port of the second network. The voltage at the output of the first network is also equal to the voltage across the input of the second network. This can be represented by using $ABCD$ matrices as

$$[ABCD] = [ABCD^x][ABCD^y] = \begin{bmatrix} A^x & B^x \\ C^x & D^x \end{bmatrix} \begin{bmatrix} A^y & B^y \\ C^y & D^y \end{bmatrix}$$

$$= \begin{bmatrix} A^x A^y + B^x C^y & A^x B^y + B^x D^y \\ C^x A^y + D^x C^y & C^x B^y + D^x D^y \end{bmatrix} \tag{1.29}$$

Example

Consider the RF amplifier given in Figure 1.6. It has feedback network for stability, input, and output matching networks. The transistor used is NPN bipolar junction transistor (BJT) and its characteristic parameters are given by $r_{BE} = 400\ \Omega$, $r_{CE} = 70\ k\Omega$, $C_{BE} = 15$ pF, $C_{BC} = 2$ pF, and $g_m = 0.2$ S. Find the voltage and current gain of this amplifier when $L = 2$ nH, $C = 12$ pF, $l = 5$ cm, and $v_p = 0.65$ c.

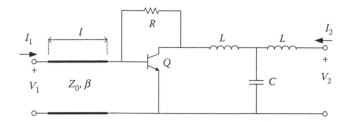

FIGURE 1.6 RF amplifier analysis by network parameters.

SOLUTION

The high-frequency characteristics of the transistor is modeled using the hybrid parameters given by

$$h_{11} = h_{ie} = \frac{r_{BE}}{1 + j\omega(C_{BE} + C_{BC})r_{BE}} \tag{1.30}$$

$$h_{12} = h_{re} = \frac{j\omega C_{BC}r_{BE}}{1 + j\omega(C_{BE} + C_{BC})r_{BE}} \tag{1.31}$$

$$h_{21} = h_{fe} = \frac{r_{BE}(g_m - j\omega C_{BC})}{1 + j\omega(C_{BE} + C_{BC})r_{BE}} \tag{1.32}$$

$$h_{22} = h_{oe} = \frac{[1 + j\omega(C_{BE} + C_{BC})r_{BE}] + [(1 + r_{BE}g_m + j\omega C_{BE}r_{BE})]r_{CE}}{1 + j\omega(C_{BE} + C_{BC})r_{BE}} \tag{1.33}$$

The amplifier network shown in Figure 1.6 is a combination of four networks that are connected in parallel and cascade. The overall network first has to be partitioned for analysis. This can be demonstrated as shown in Figure 1.7.

In the partitioned amplifier circuit, networks N_2 and N_3 are connected in parallel as shown in Figure 1.8. Then, the parallel-connected network, Y, can be represented by the admittance matrix. The admittance matrix of network 3 is

$$Y^y = \begin{bmatrix} \dfrac{1}{R} & -\dfrac{1}{R} \\ -\dfrac{1}{R} & \dfrac{1}{R} \end{bmatrix} \tag{1.34}$$

The admittance matrix for the transistor can now be obtained by using the network conversion Table 1.1 since hybrid parameters for it are available. From Table 1.1

$$Y^x = \begin{bmatrix} \dfrac{1}{h_{11}} & -\dfrac{h_{12}}{h_{11}} \\ \dfrac{h_{21}}{h_{11}} & \dfrac{\Delta h}{h_{11}} \end{bmatrix} \tag{1.35}$$

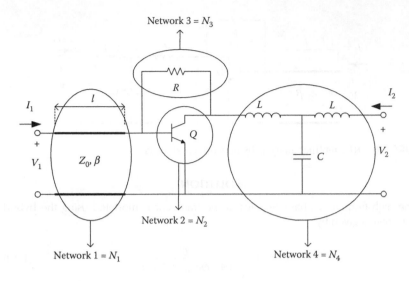

FIGURE 1.7   Partition of the amplifier circuit for network analysis.

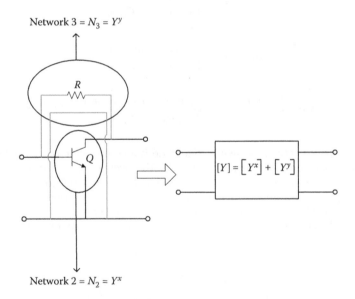

FIGURE 1.8   Illustration of the parallel connection between networks 2 and 3.

Then, the overall admittance matrix is found as

$$
[Y] = \left[Y^x \right] + \left[Y^y \right] =
\begin{bmatrix}
\dfrac{1}{R} + \dfrac{1}{h_{11}} & -\dfrac{1}{R} - \dfrac{h_{12}}{h_{11}} \\[2ex]
-\dfrac{1}{R} + \dfrac{h_{21}}{h_{11}} & \dfrac{1}{R} + \dfrac{\Delta h}{h_{11}}
\end{bmatrix}
\tag{1.36}
$$

where Δ is used for the determinant of the corresponding matrix. At this point, it is now more clear that networks 1, Y, and 4 are cascaded. We need to determine the $ABCD$ matrix of each network in this circuit as shown in Figure 1.9. The first step is then to convert the admittance matrix in Equation 1.36 to the $ABCD$ parameter using the conversion table. The conversion table gives the relation as

$$ABCD^Y = \begin{bmatrix} \dfrac{Y_{22}}{Y_{21}} & -\dfrac{1}{Y_{21}} \\ \dfrac{\Delta Y}{Y_{21}} & \dfrac{Y_{11}}{Y_{21}} \end{bmatrix} \tag{1.37}$$

$ABCD$ matrices for the transmission line network 1 and the matching network 4 are obtained as

$$ABCD^{N_1} = \begin{bmatrix} \cos(\beta l) & jZ_0 \sin(\beta l) \\ \dfrac{j\sin(\beta l)}{Z_0} & \cos(\beta l) \end{bmatrix} \tag{1.38}$$

$$ABCD^{N_4} = \begin{bmatrix} 1 - \omega^2 LC & j\omega L(2 - \omega^2 LC) \\ j\omega C & 1 - \omega^2 LC \end{bmatrix} \tag{1.39}$$

The overall $ABCD$ parameter of the combined network shown in Figure 1.9 is

$$ABCD = ABCD^{N_1}(ABCD^Y)ABCD^{N_4} \tag{1.40}$$

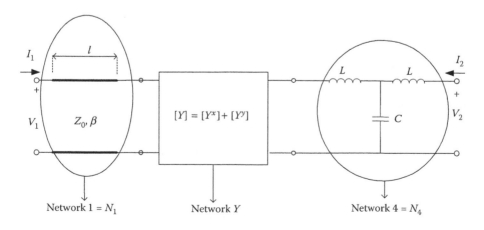

FIGURE 1.9 Cascade connection of the final circuit.

or

$$ABCD = \begin{bmatrix} \cos(\beta l) & jZ_0\sin(\beta l) \\ \dfrac{j\sin(\beta l)}{Z_0} & \cos(\beta l) \end{bmatrix} \begin{bmatrix} \dfrac{Y_{22}}{Y_{21}} & -\dfrac{1}{Y_{21}} \\ \dfrac{\Delta Y}{Y_{21}} & \dfrac{Y_{11}}{Y_{21}} \end{bmatrix} \begin{bmatrix} 1-\omega^2 LC & j\omega L(2-\omega^2 LC) \\ j\omega C & 1-\omega^2 LC \end{bmatrix}$$

$$(1.41)$$

The voltage and current gains from *ABCD* parameters are found using

$$V_{Gain} = 20\log\left(\left|\frac{1}{A}\right|\right)(\text{dB})$$ (1.42)

$$I_{Gain} = 20\log\left(\left|\frac{1}{D}\right|\right)(\text{dB})$$ (1.43)

The MATLAB® script of the program is shown below.

MATLAB Script for the Network Analysis of RF Amplifier

```
Zo = 50;
l = 0.05;
L = 2e-9;
C = 12e-12;
rbe = 400;
rce = 70e3;
Cbe = 15e-12;
Cbc = 2e-12;
gm = 0.2;
VGain = zeros(5,150);
IGain = zeros(5,150);
freq = zeros(1,150);
R = [200 300 500 1000 10000];

for i = 1:5
for t = 1:150;

f = 10^((t + 20)/20);
freq(t) = f;
lambda = 0.65*3e8/(f);
bet = (2*pi)/lambda;
w = 2*pi*f;
N1 = [cos(bet*l) 1j*Zo*sin(bet*l);1j*(1/Zo)*sin(bet*l) cos(bet*l)];
Y1 = [1/R(i) -1/R(i);-1/R(i) 1/R(i)];
k = (1 + 1j*w*rbe*(Cbc + Cbe));
h = [(rbe/k) (1j*w*rbe*Cbc)/k;(rbe.*(gm-1j*w*Cbc))/k ((1/rce)
+ (1j*w*Cbc*(1 + gm*rbe + 1j*w*Cbe*rbe)/k))];
Y2 = [1/h(1,1) -h(1,2)/h(1,1);h(2,1)/h(1,1) det(h)/h(1,1)];
Y = Y1 + Y2;
N23 = [-Y(2,2)/Y(2,1) -1/Y(2,1);(det(Y)/Y(2,1)) -Y(1,1)/Y(2,1)];
```

```
N4 = [(1-(w^2)*L*C) (2j*w*L-1j*(w^3)*L^2*C);1j*(w*C) (1-(w^2)*L*C)];
NT = N1*N23*N4;
VGain(i,t) = 20*log10(abs(1/NT(1,1)));
IGain(i,t) = 20*log10(abs(-1/NT(2,2)));

end
end

figure
semilogx(freq,(IGain))
axis([10^4 10^9 20 50]);
ylabel('I_{Gain} (I_2/I_1) (dB)');
xlabel('Freq (Hz)');
legend('R = 200Ohm','R = 300Ohm','R = 500Ohm','R = 1000Ohm',' R = 10000Ohm')
figure
semilogx(freq,(VGain))
axis([10^4 10^9 20 80]);
ylabel('V_{Gain} (V_2/V_1) (dB)');
xlabel('Freq (Hz)');
legend('R = 200Ohm','R = 300Ohm','R = 500Ohm','R = 1000Ohm',' R = 10000Ohm')
```

MATLAB script has been written to obtain the voltage and current gains. The script that can be used for analysis of any other similarly constructed amplifier network is given for reference. Voltage and current gains obtained by MATLAB versus various feedback resistor values and frequency are shown in Figures 1.10 and 1.11. This type of analysis gives the designer the ability to study the effect of several parameters on the output response in an amplifier circuit including feedback-matching networks and parameters of the transistor.

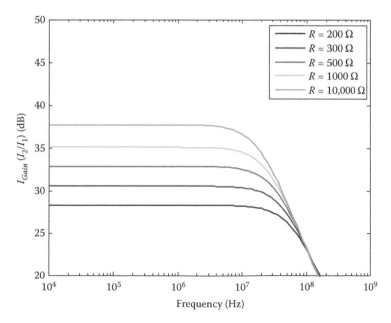

FIGURE 1.10 Current gain of the RF amplifier versus feedback resistor values and frequency.

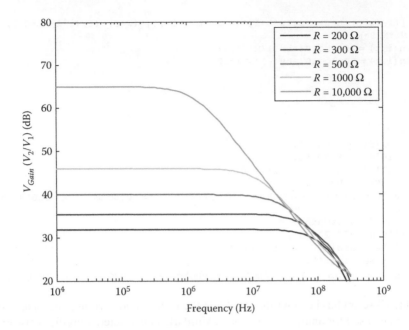

FIGURE 1.11   Voltage gain of the RF amplifier versus feedback resistor values and frequency-matching networks and parameters of the transistor.

1.4   *S*-SCATTERING PARAMETERS

Scattering parameters are used to characterize RF/microwave devices and components at high frequencies. Specifically, they are used to define the return loss and insertion loss of components or devices. S parameters can be analyzed using a two-port network shown in Figure 1.12.

In Figure 1.12, the incident-normalized wave a_n and a reflected-normalized wave b_n are related through S parameters as

$$\begin{bmatrix} b_1 \\ b_2 \end{bmatrix} = \begin{bmatrix} S_{11} & S_{12} \\ S_{21} & S_{22} \end{bmatrix} \begin{bmatrix} a_1 \\ a_2 \end{bmatrix} \tag{1.44}$$

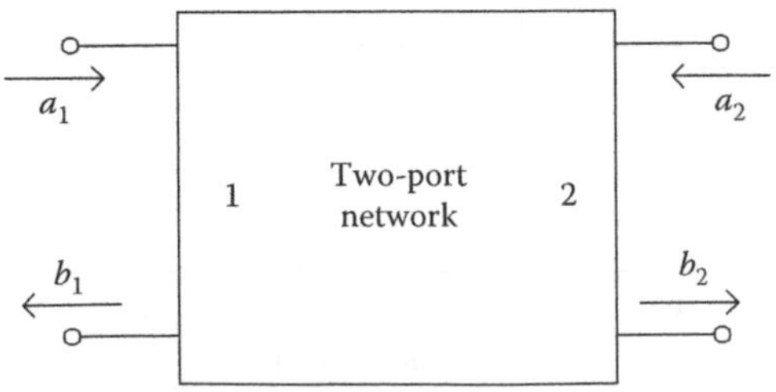

FIGURE 1.12   *S* parameters for two-port networks.

S parameters in Equation 1.44 are described by the following equations:

$$S_{11} = \left.\frac{b_1}{a_1}\right|_{a_2=0} \quad S_{12} = \left.\frac{b_1}{a_2}\right|_{a_1=0}$$
$$S_{21} = \left.\frac{b_2}{a_1}\right|_{a_2=0} \quad S_{22} = \left.\frac{b_2}{a_2}\right|_{a_1=0}$$

(1.45)

In Equation 1.45, S_{11} is the reflection coefficient at the input, S_{22} is the reflection coefficient at the output, S_{21} is the forward transmission gain, and S_{12} is the reverse transmission gain. As seen from Equation 1.45, S parameters are found when $a_n = 0$ with no reflection at other ports. This is only possible by matching all the ports except the measurement port. The insertion loss and return loss in terms of S parameters are defined as

$$\text{Insertion loss (dB)} = IL\,(\text{dB}) = 20\log\big(|S_{ij}|\big), \quad i \neq j$$

(1.46)

$$\text{Return loss (dB)} = RL\,(\text{dB}) = 20\log\big(|S_{ii}|\big)$$

(1.47)

Another important parameter that can be defined using S parameters is the voltage standing wave ratio, VSWR. For instance, VSWR at the input, port 1, is found from

$$\text{VSWR} = \frac{1-|S_{11}|}{1+|S_{11}|}$$

(1.48)

The two-port network is reciprocal if

$$S_{21} = S_{12}$$

(1.49)

It can be shown that a network is reciprocal if its scattering matrix is equal to its transpose. This is represented for two-port networks as

$$[S] = [S]^t$$

(1.50)

or

$$\begin{bmatrix} S_{11} & S_{12} \\ S_{21} & S_{22} \end{bmatrix}^t = \begin{bmatrix} S_{11} & S_{21} \\ S_{12} & S_{22} \end{bmatrix}$$

(1.51)

S parameters can be used to characterize lossless networks as

$$[S]^t [S]^* = [U]$$

(1.52)

where * defines the complex conjugate of a matrix and U is the unitary matrix defined by

$$[U] = \begin{bmatrix} 1 & 0 \\ 0 & 1 \end{bmatrix} \tag{1.53}$$

Equation 1.52 can be applied for two-port networks as

$$[S]^t [S]^* = \begin{bmatrix} \left(|S_{11}|^2 + |S_{21}|^2\right) & \left(S_{11}S_{12}^* + S_{21}S_{22}^*\right) \\ \left(S_{12}S_{11}^* + S_{22}S_{21}^*\right) & \left(|S_{12}|^2 + |S_{22}|^2\right) \end{bmatrix} = \begin{bmatrix} 1 & 0 \\ 0 & 1 \end{bmatrix} \tag{1.54}$$

We can further show that if a network is lossless and reciprocal, it satisfies

$$|S_{11}|^2 + |S_{21}|^2 = 1 \tag{1.55}$$

$$S_{11}S_{12}^* + S_{21}S_{22}^* = 0 \tag{1.56}$$

Scattering parameters can be generalized for multiport network as shown in Figure 1.13 as

$$\begin{bmatrix} b_1 \\ b_2 \\ b_3 \\ b_4 \\ b_5 \end{bmatrix} = \begin{bmatrix} S_{11} & S_{12} & S_{13} & S_{14} & S_{15} \\ S_{21} & S_{22} & S_{23} & S_{24} & S_{25} \\ S_{31} & S_{32} & S_{33} & S_{34} & S_{35} \\ S_{41} & S_{42} & S_{43} & S_{44} & S_{45} \\ S_{51} & S_{52} & S_{53} & S_{54} & S_{55} \end{bmatrix} \begin{bmatrix} a_1 \\ a_2 \\ a_3 \\ a_4 \\ a_5 \end{bmatrix} \tag{1.57}$$

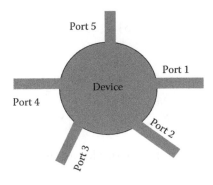

FIGURE 1.13 S parameters for multiport networks.

Example

Find the characteristic impedance of the T-network shown in Figure 1.3 to have no return loss at its input port.

SOLUTION

Scattering parameters for T-network are found from Equation 1.45. From Equation 1.45, S_{11} is equal to

$$S_{11} = \frac{b_1}{a_1}\bigg|_{a_2=0} = \frac{Z_{in} - Z_o}{Z_{in} + Z_o} \tag{1.58}$$

where

$$Z_{in} = Z_A + \left[\frac{Z_C(Z_B + Z_o)}{Z_C + (Z_B + Z_o)}\right] \tag{1.59}$$

No return loss is possible when $S_{11} = 0$. This can be satisfied from Equation 1.58 when

$$Z_o = Z_{in} = Z_A + \left[\frac{Z_C(Z_B + Z_o)}{Z_C + (Z_B + Z_o)}\right] \tag{1.60}$$

REFERENCES

1. G.L. Matthaei, L. Young, and E.M.T. Jones, *Microwave Filters, Impedance Matching Networks, and Coupling Structures*, Artech House, Dedham, MA, 1980.
2. D.A. Frickey, Conversions between S, Z, Y, h, ABCD and T parameters which are valid for complex source and load impedances, *IEEE Transactions on Microwave Theory and Techniques*, 42(2), 205–211, 1994.
3. F.F. Kuo, *Network Analysis and Synthesis*, Wiley, New York, 1996.
4. Y. Kuznetsov, A. Baev, P. Lorenz, and P. Russer, Network oriented modeling of passive microwave structures, *EUROCON 2007: The International Conference on Computer as a Tool*, Warsaw, pp. 10–14, September 9–12, 2007.
5. E.A. Guillemin, *Synthesis of Passive Networks*, Wiley, New York, 1957.

2 MF-UHF Inductor Design Techniques

2.1 INTRODUCTION

If a current flows through a wire wound, a flux is produced through each turn as a result of magnetic flux density as shown in Figure 2.1. The relation between the flux density and the flux through each turn can be represented as

$$\Psi = \int \bar{B} \cdot d\bar{s} \qquad (2.1)$$

If there are N turns, then we define the flux linkage as

$$\lambda = N\Psi = N\int \bar{B} \cdot d\bar{s} \qquad (2.2)$$

Inductance is defined as the ratio of flux linkage to the current flowing through the windings and defined by

$$L = \frac{\lambda}{I} = \frac{N\Psi}{I} \qquad (2.3)$$

The inductance defined by Equation 2.3 is also known as self-inductance of the core that is formed by the windings. The core can be an air core or a magnetic core.

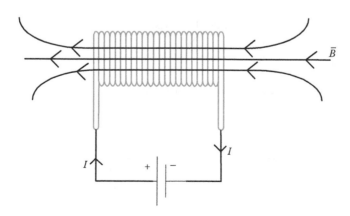

FIGURE 2.1 Flux through each turn.

2.2 AIR CORE INDUCTOR DESIGN AND EXAMPLES

Inductors can be formed as air core inductors or magnetic core inductors depending on the application. When air core inductors are formed through windings and operated at high frequency, the inductor presents high-frequency characteristics. This includes winding resistance and distributed capacitance effects between each turn as shown in Figure 2.2.

The high-frequency model of the inductor can be illustrated with its equivalent circuit shown in Figure 2.3.

As a result, the inductor will act as an inductor up to a certain frequency, then get into resonance, and exhibit capacitive effects after the resonance frequency. The equation giving these characteristics can be written as

$$Z = \frac{(j\omega L + R)(1/j\omega C_s)}{(j\omega L + R) + (1/j\omega C_s)} = \frac{R}{(1 - \omega^2 LC_s)^2 + (\omega RC_s)^2} + j\frac{\omega(L - R^2 C_s) - \omega^3 L^2 C_s}{(1 - \omega^2 LC_s)^2 + (\omega RC_s)^2}$$

(2.4)

Equation 2.4 can also be expressed as

$$Z = R_s + jX_s \tag{2.5}$$

where

$$R_s = \frac{R}{\left(1 - \omega^2 LC_s\right)^2 + \left(\omega RC_s\right)^2} \tag{2.6a}$$

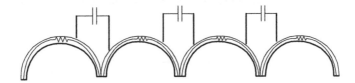

FIGURE 2.2 High-frequency effects of RF inductor.

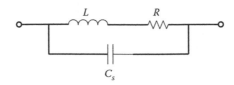

FIGURE 2.3 High-frequency model of RF air core inductor.

FIGURE 2.4 Equivalent series circuit.

and

$$X_s = \frac{\omega\left(L - R^2 C_s\right) - \omega^3 L^2 C_s}{\left(1 - \omega^2 L C_s\right)^2 + \left(\omega R C_s\right)^2} \tag{2.6b}$$

This is in fact the representation of the equivalent series circuit shown in Figure 2.4. The resonance frequency is found when $X_s = 0$ as

$$f_r = \frac{1}{2\pi}\sqrt{\frac{L - R^2 C_s}{L^2 C_s}} \tag{2.7}$$

The quality factor is obtained from

$$Q = \frac{|X_s|}{R_s} \tag{2.8}$$

R_s in Figure 2.4 is the series resistance of the inductor and includes the distributed resistance effect of the wire. It is calculated from

$$R = \frac{l_w}{\sigma A} \tag{2.9}$$

C_s in Figure 2.3 is the capacitance including the effects of distributed capacitance of the inductor and given by

$$C_s = \frac{2\pi\varepsilon_0 da N^2}{l_w} \tag{2.10}$$

Example 2.1

Obtain the high-frequency characteristics of the inductor using the model given in Figure 2.3 when $L = 10$ [nH], $C_s = 1$ [pF], and $R = 75$ [Ω].

SOLUTION

The MATLAB® script of the program is given as

MATLAB Script: High-Frequency Characteristics of Inductor

```
%High Frequency Model of Inductor, Ex 1
f = linspace(1,6*10^9);
```

```
Lext = 10*10^(-9);
Ca = 1*10^(-12);
R = 75;
w = 2*pi.*f;
Z = (i*w.*Lext + R)./((i*w.*Lext + R).*(i*w.*Ca) + 1);
magZ = abs(Z);
res = (1/(2*pi))*sqrt((Lext-R^2*Ca)/((Lext^2).*Ca))
plot(f,magZ)
title('High Frequency Inductor response')
xlabel('f (Hz)')
ylabel('Magnitude of Z')
```

The response is shown in Figure 2.5.

In practice, inductors can be implemented as air core or toroidal inductors as illustrated in Figure 2.6.

For an air core selonoidal inductor given in Figure 2.6a, the inductance can be calculated using the relation

$$L = \frac{d^2 N^2}{18d + 40l} \quad [\mu H] \tag{2.11}$$

In this equation, L is given as inductance in [μH], d is the coil inner diameter in inches (in), l is the coil length in inches (in), and N is the number of turns of the coil.

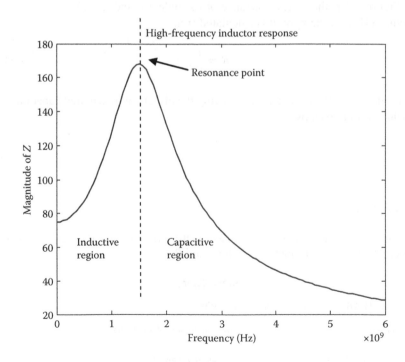

FIGURE 2.5 High-frequency characteristics response.

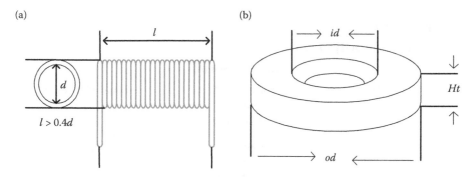

FIGURE 2.6 (a) Air core inductor. (b) Toroidal inductor.

The formula given in Equation 2.11 can be extended to include the spacing between each turn of the air coil inductor. Then, Equation 2.11 can be modified as

$$L = \frac{d^2N^2}{18d + 40(Na + (N-1)s)} \quad [\mu H] \tag{2.12}$$

In Equation 2.12, a represents the wire diameter in inches, and s represents the spacing in inches between each turn. The wire dimensions are given in Table 2.1.

TABLE 2.1
Wire Sizes and Their Physical Dimensions

AWG	Diameter (inch)	Diameter (mm)	AWG	Diameter (inch)	Diameter (mm)
2	0.2576	6.544	22	0.0253	0.644
3	0.2294	5.827	23	0.0226	0.573
4	0.2043	5.189	24	0.0201	0.511
5	0.1819	4.621	25	0.0179	0.455
6	0.162	4.115	26	0.0159	0.405
7	0.1443	3.665	27	0.0142	0.361
8	0.1285	3.264	28	0.0126	0.321
9	0.1144	2.906	29	0.0113	0.286
10	0.1019	2.588	30	0.01	0.255
11	0.0907	2.305	31	0.00893	0.227
12	0.0808	2.053	32	0.00795	0.202
13	0.072	1.828	33	0.00708	0.18
14	0.0641	1.628	34	0.0063	0.16
15	0.0571	1.45	35	0.00561	0.143
16	0.0508	1.291	36	0.005	0.127
17	0.0453	1.15	37	0.00445	0.113
18	0.0403	1.024	38	0.00397	0.101
19	0.0359	0.912	39	0.00353	0.0897
20	0.032	0.812	40	0.00314	0.0799

TABLE 2.2
Current Capacity of Wires

Wire Size (AWG)	Derated Current (A)	
	Solid Wire (A)	Stranded Wire (A)
40	0.226134423	0.113067211
38	0.314649454	0.157324727
36	0.437811623	0.218905811
34	0.609182742	0.304591371
32	0.847633078	0.423816539
30	1.179419221	0.58970961
28	1.641075289	0.820537644
26	2.283435826	1.141717913
24	3.177233372	1.588616686
22	4.420887061	2.21044353
20	6.151339898	3.075669949
18	8.559138024	4.279569012
16	11.90941241	5.954706204
14	16.57107334	8.28553667
12	23.05743241	11.5287162
10	32.08272502	16.04136251
8	44.64075732	22.32037866
6	62.11433765	31.05716883
4	86.42754229	43.21377115
2	120.2575822	60.12879111
0	167.3296	83.6648

The current capacity of each solid and stranded wires is calculated based on MIL-STD-975, including the effect of derating. This is illustrated in Table 2.2 and Figure 2.7.

Example 2.2

What is the required number of turns and length of the inductor when 12AWG is used to form an air inductor with a 0.5 in rod for winding to form 330 nH inductance? Obtain the high-frequency characteristics of this inductor, identify its resonant frequency, and find the quality factor.

SOLUTION

From Equation 2.12 with neglecting the spacing between each turn

$$L = \frac{d^2 N^2}{18d + 40(Na)} \quad [\mu H] \tag{2.13}$$

where $l_{ind} = Na$ is used to find the length of the inductor. This leads to the following quadratic equation:

$$0.25N^2 - 1.066N - 2.97 = 0 \tag{2.14}$$

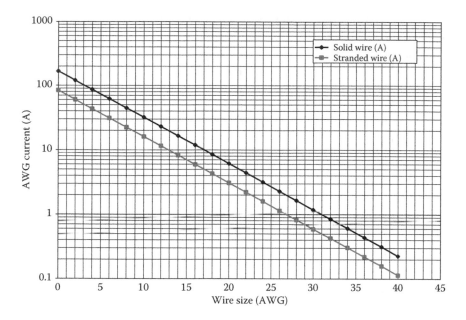

FIGURE 2.7 Plot of the current capacity of solid and stranded wires.

The solution for Equation 2.14 is satisfied when $N = 6.18$. N has to be a positive integer. The value of inductance and length for $N = 6$ and $N = 7$ using Equation 2.13 are given below:

$$N = 6, \quad L = 0.317 \text{ [μH]}, \quad l_{ind} = 0.4848 \text{ [in]}$$
$$N = 7, \quad L = 0.387 \text{ [μH]}, \quad l_{ind} = 0.5656 \text{ [in]} \tag{2.15}$$

The designer has to choose the number of turns based on the design to have minimal impact. We choose the number of turns to be equal to $N = 7$ for high-frequency characteristic calculation. Then, the length of the air core inductor is

$$l_{ind} = Na = 7\,(0.0808) = 0.5656 \text{ [in]} = 0.01144 \text{ [m]}$$

Since $\sigma = 5.96 \times 10^7$ [S/m], $l_w = 2\pi rN = 0.045$ [m], the cross-sectional area of the wire can be found to be $A = \pi(2.053 \times 10^{-3}/2)^2 = 3.31 \times 10^{-6}$ [m²]. Substituting these into Equations 2.9 and 2.10 gives

$$R = \frac{l_w}{\sigma A} = \frac{0.045}{(5.96 \times 10^7)(3.31 \times 10^{-6})} = 0.228 \text{ [mΩ]}$$

and

$$C_s = \frac{2\pi \varepsilon_0 daN^2}{l_w} = \frac{2\pi(8.85 \times 10^{-12})(0.0127)(2.053 \times 10^{-3})49}{0.045} = 1.578 \text{ [pF]}$$

The high-frequency response of the inductor is illustrated in Figure 2.8.
The plot of quality factor versus frequency for this inductor is given in Figure 2.9.

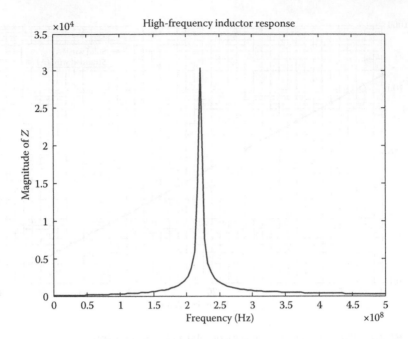

FIGURE 2.8 High-frequency characteristics response of the air core inductor.

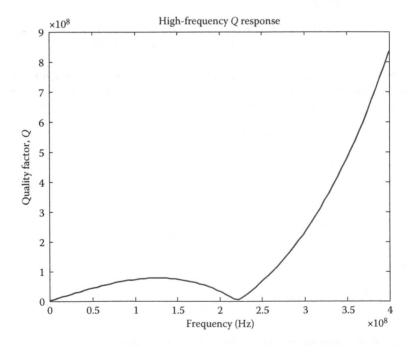

FIGURE 2.9 Quality factor of the air core inductor versus frequency.

The resonant frequency is found from Equation 2.7 as

$$f_r = \frac{1}{2\pi}\sqrt{\frac{L - R^2 C_s}{L^2 C_s}} = \frac{1}{2\pi}\sqrt{\frac{330 \times 10^{-9} - (0.228 \times 10^{-3})^2 (1.578 \times 10^{-12})}{(330 \times 10^{-9})^2 (1.578 \times 10^{-12})}}$$

$$= 2.2 \times 10^8 \; [Hz]$$

The resonant frequency calculated above matched with the results illustrated in Figures 2.8 and 2.9. Air core inductors inherently have very high Q factors. At the resonance frequency, the quality factor is zero as expected.

2.3 TOROIDAL INDUCTOR DESIGN AND EXAMPLES

In several RF applications, it maybe required to have larger inductance values in space-restricted areas. One solution to increase the inductance value for an air core inductor is to increase the number of turns. However, this increases the size of the air core inductor. This challenge can be overcome by using toroidal cores. One of the other advantages of using toroidal cores is keeping the flux within the core as shown in Figure 2.10. This provides self-shielding. In the air core inductor design, air is used as a nonmagnetic material to wind the wire around it. When air is replaced with a magnetic material such as the toroidal core, the inductance of the formed inductor can be calculated using

$$L = \frac{4\pi N^2 \mu_i A_{Tc}}{l_e} \; [nH] \tag{2.16a}$$

In Equation 2.16a, L is the inductance in nanohenries (nH), N is the number of turns, μ_i is the initial permeability, A_{Tc} is the total cross-sectional area of the core in centimeter^2, and l_e is the effective length of the core in centimeter. The effective length of the core l_e is defined as

$$l_e = \frac{\pi(od - id)}{\ln(od/id)} \; [cm] \tag{2.16b}$$

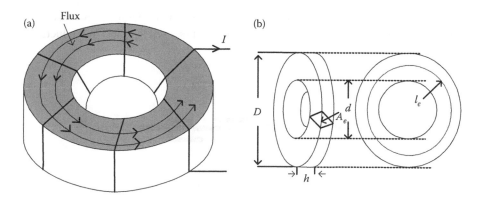

FIGURE 2.10 (a) Magnetic flux in the toroidal core. (b) Geometry of the toroid.

where *od* is the outside diameter and *id* is the inside diameter of the core in centimeter. The total cross-sectional area of the core, A_{Tc}, is defined as

$$A_{Tc} = \frac{1}{2}(od - id) \times h \times n \ [\text{cm}^2]$$ (2.16c)

h refers to the thickness of the core in centimeter and *n* is used to define the number of stacked cores.

It is not uncommon to have the information about the inductance index of the core on its data sheet. If the inductance index is given, then Equation 2.16a can be modified as

$$L = N^2 A_L \ [\text{nH}]$$ (2.17)

where A_L is the inductance index in nanohenries/turn². Once the number of turns is identified for the required inductance value using Equations 2.16 and 2.17, the wire size has to be determined. The turn for toroidal core is defined as the pass through the core as shown in Figure 2.11.

The wire maximum diameter can then be determined using

$$d = \frac{2\pi(id)}{N + \pi} \ [\text{in}]$$ (2.18)

d is known as the diameter of the wire that is used in inches, *id* is the inner diameter of the core shown in Figure 2.6b, and *N* is the number of turns.

The length of the wire that is used to wind the core shown in Figure 2.12 can be approximated as

$$l_w = n \times \pi \sqrt{[N(2r + od - id)]^2 + (od + id)^2} \ [\text{cm}]$$ (2.19)

where *r* is the radius of the wire, *w* is the width of the core, *h* is the thickness of the core, and *n* is the number of stacked cores. As the wire diameter increases, the number of turns, *N*, decreases. This information can be used to adjust the inductance

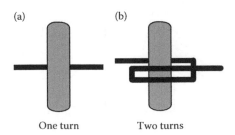

FIGURE 2.11 Implementation of turn for toroidal cores. (a) One turn. (b) Two turns.

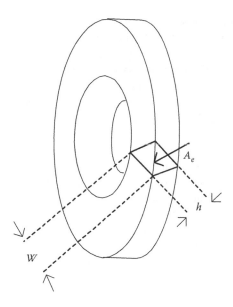

FIGURE 2.12   Toroidal core wire length calculation.

value in Equation 2.16a to accommodate the current flow needed. The current rating of solid and stranded wires can be found from the plot given in Figure 2.7. The maximum operational flux density for toroidal core is calculated from

$$B_{op} = \frac{V_{rms} \times 10^8}{4.44\, fNA_{T_c}} \quad [\text{Gauss}] \tag{2.20}$$

In Equation 2.20, B_{op} is the magnetic flux density in Gauss, V_{rms} is the maximum *rms* voltage across the inductor in volts, f is the frequency in Hertz, N is the number of turns, and A_{T_c} is the total cross-sectional area of the core in centimeter². The proper design of the toroidal core inductor requires the operational voltage of the inductor and the required inductance value. This helps to identify the right material for the inductor design to prevent saturation.

It is then the designer's task to use Equations 2.16a and 2.20 to determine the core, the number of turns, and then the reliable operational flux density.

The core material carries great importance in the design of toroidal-type inductors. The main types of magnetic core materials used in the inductor design are classified as powdered iron cores and ferrites. The important difference that ties with the application of the specific type of core depends on its core loss and saturation. The core saturation level can be increased by introducing an air gap within the flux path in the core. The air gap for powdered iron cores is naturally distributed along the flux path. As a result, they inherently have higher saturation levels in comparison to ferrite cores. This results in higher levels of current flow through the inductor before it saturates. The air gap for ferrites has to be created physically during the manufacturing process. This is the reason why the saturation takes place gradually for powdered iron cores whereas it happens at once for ferrite cores. When saturation takes place, it is permanent for ferrite cores and temporary for powdered iron cores. The main material type used for powdered iron cores are carbonyl irons and hydrogen-reduced irons. The permeability of carbonyl-type powdered iron cores are less than hydrogen-reduced types and can get values up to 35. Hydrogen-reduced powdered iron cores can have permeability values up to 90; however, their Qs are less than carbonyl types. During the manufacturing process of iron cores, material properties can be changed to obtain some critical parameters such as higher permeability and flux density. The effect of changes on material properties during the manufacturing of iron cores is illustrated in Table 2.3.

The losses in toroidal inductor can be categorized under core losses and copper losses [2]. Core losses are due to hysteresis loss and eddy current loss whereas copper loss happens due to windings. The varying magnetic field in a conductive material induces eddy current loss as in iron powder cores. Eddy current loss can be reduced by using higher-resistivity material. Hysteresis loss is due to hysteresis phenomenon in magnetic materials where some of the magnetization is converted to heat and as a result it is considered as loss. Winding losses are the copper losses that happen due to current flow in conductive wires, which have certain resistance values. Litz wires can be used to reduce the winding losses. Litz wires consist of stranded wires that have diameter less than the skin depth, which is given by the following relation:

$$\delta = \sqrt{\frac{\rho}{\pi f \mu_0 \mu_r}} \qquad (2.21)$$

where ρ is the resistivity of the wire conductor, f is the operational frequency, μ_0 is the free space permeability, and μ_r is the relative permeability of the wire conductor.

TABLE 2.3

Changes in Material Properties of Iron Cores during Manufacturing

	Increasing Particle Size	Addition of Lubricant	Addition of Binder	Increasing Compaction Pressure	Heat Treatment
Permeability	↑	↓	↓	↑	↑
Maximum flux density	↑	↓	↓	↑	↑
Coercivity	↓	→	→	↑	↓
Resistivity	↓	↑	↑	↓	↓
Thermal conductivity	↑	↓	↓	↑	↑
Strength	↓	↓	↑	↑↓[a]	↑[a]

[a] The direction changes with the inclusion of the binder.

TABLE 2.4

Frequency Response of Powdered Iron Cores

MIX	COLOR	
26	Yellow/white	
3	Gray	
15	Red/white	
1	Blue	
2	Red	
7	White	
6	Yellow	
10	Black	
12	Green/white	
17	Blue/yellow	
0	Tan	

0.05 0.10 0.50 1.0 3.0 5.0 10 30 50 100 200 300

FREQUENCY (MHz)

Core manufacturers [11–13] use color code and number system for various materials for identification. For instance, Table 2.4 illustrates the material composition, the color code, and the recommended operational frequency for different types of cores.

Table 2.5 illustrates the material properties and permeability of cores based on the color codes and number system. Table 2.6 gives the application detail for each type of core, which is identified with the number and color code systems. Each toroid core comes in different size. There is a number system that identifies the physical dimensions of each core. The figure illustrating the physical dimensions of each core based on the number system is given in Table 2.7. Table 2.8 shows the inductance index, A_L, values using the code system with physical dimensions. Ferrites

TABLE 2.5

Permeability and Material Information

Type	μ	Color Code	Material
26	75	Yellow/white	Hydrogen reduced
3	35	Gray	Carbonyl HP
15	25	Red/white	Carbonyl GS6
1	20	Blue	Carbonyl C
2	10	Red	Carbonyl E
7	9	White	Carbonyl TH
6	8	Yellow	Carbonyl SF
10	6	Black	Powdered iron SF
12	3	Green/white	Synthetic oxide
17	3	Blue/yellow	Carbonyl
0	1	Tan	Phenolic

TABLE 2.6
Detailed Material Information

Type	Uses
26	High permeability, used in EMI filters, DC chokes, and switched DC power supplies
3	High-Q coils and transformers between 50 and 500 kHz
15	Good Q and high stability. Commonly used in AM BCB and 160 m amateur applications
1	High volume resistivity. Used for lower-frequency applications
2	High volume resistivity. Commonly used for inductors and transformers in the 3–30 MHz HF bands
7	Used for HF and low-end VHF inductors and transformers
6	Offers higher Q between 30 and 50 MHz, but is used for HF and low VHF band inductors and transformers
10	Good Q and high stability for use in inductors and transformers between 40 and 100 MHz
12	Good Q but only moderate stability for inductors and transformers between 50 and 100 MHz
17	Similar to Type 12, but has better temperature stability and lower Q
0	High-Q applications above 200 MHz. The actual inductance is more sensitive to the winding technique than other types

TABLE 2.7
Dimensions of Cores Used with the Color Codes

Core Size	O.D. (in)	O.D. (mm)	I.D. (in)	I.D. (mm)	H (in)	H (mm)
T-12	0.125	3.175	0.062	1.575	0.05	1.270
T-16	0.160	4.064	0.078	1.981	0.06	1.524
T-20	0.200	5.080	0.088	2.235	0.07	1.778
T-25	0.250	6.350	0.12	3.048	0.096	2.438
T-30	0.307	7.798	0.151	3.835	0.128	3.251
T-37	0.375	9.525	0.205	5.207	0.128	3.251
T-44	0.440	11.176	0.229	5.817	0.159	4.039
T-50	0.500	12.700	0.300	7.620	0.190	4.826
T-68	0.690	17.526	0.370	9.398	0.190	4.826
T-80	0.795	20.193	0.495	12.573	0.250	6.350
T-94	0.942	23.927	0.560	14.224	0.312	7.925
T-106	1.060	26.924	0.570	14.478	0.437	11.100
T-130	1.300	33.020	0.780	19.812	0.437	11.100
T-157	1.570	39.878	0.950	24.130	0.570	14.478
T-184	1.840	46.736	0.950	24.130	0.710	18.034
T-200	2.000	50.800	1.250	31.750	0.550	13.970
T-200A	2.000	50.800	1.250	31.750	1.000	25.400
T-225	2.250	57.150	1.400	35.560	0.550	13.970
T-225A	2.250	57.150	1.400	35.560	1.000	25.400
T-300	3.000	76.200	1.920	48.768	0.500	12.700
T-300A	3.000	76.200	1.920	48.768	1.000	25.400
T-400	4.000	101.600	2.250	57.150	0.650	16.510
T-400A	4.000	101.600	2.250	57.150	1.000	25.400
T-500	5.200	132.080	3.080	78.232	0.800	20.320

TABLE 2.8

Inductance Index, A_L Values in Color and Number Systems

Mix	26	3	15	1	2	7	6	10	12	17	0
Color	Yellow/white	Gray	Red/white	Blue	Red	White	Yellow	Black	Green/white	Blue/yellow	Tan
Material	H reduced	Carb HP	Carb GS6	Carb C	Carb E	Carb TH	Carb SF	Powdered iron SF	Syn oxide	Carb	Phenolic
Frequency (MHz)	DC-1	0.05–0.50	0.10–2	0.5–5	2–30	3–35	10–50	30–100	50–200	40–180	100–300
μ	75	35	25	20	10	9	8	6	4	4	1
Temperature coefficient (PPM/C)	825	370	190	280	95	30	35	150	170	50	0
Core Size					A_L Values						
T-12	N/A	60	50	48	20	18	17	12	7.5	7.5	3
T-16	145	61	55	44	22	N/A	19	13	8	8	3
T-20	180	76	65	52	27	24	22	16	10	10	3.5
T-25	235	100	85	70	34	29	27	19	12	12	4.5
T-30	325	140	93	85	43	37	36	25	16	16	6
T-37	275	120	90	80	40	32	30	25	15	15	4.9
T-44	360	180	160	105	52	46	42	33	18.5	18.5	6.5
T-50	320	175	135	100	49	43	40	31	18	18	6.4
T-68	420	195	180	115	57	52	47	32	21	21	7.5
T-90	450	180	170	115	55	50	45	32	22	22	8.5
T-94	590	248	200	160	84	N/A	70	58	32	N/A	10.6
T-106	900	450	345	325	135	133	116	N/A	N/A	N/A	19
T-130	785	350	250	200	110	103	96	N/A	N/A	N/A	15
T-157	870	420	360	320	140	N/A	115	N/A	N/A	N/A	N/A
T-184	1640	720	N/A	500	240	N/A	195	N/A	N/A	N/A	N/A
T-200	895	425	N/A	250	120	105	100	N/A	N/A	N/A	N/A

Note: H, hydrogen; Carb, carbonyl; Syn, synthetic.

are ceramic materials, which are mixtures of iron oxide with oxides or carbonates of either manganese and zinc or nickel and zinc. The nickel–zinc cores have a high volume resistivity, good stability, and high Q factors. The common ferrite cores used in RF applications have permeability ranges from 40 to 850. The material properties of the ferrite materials commonly used in RF applications are given in Table 2.9.

The composition used for each material type is given in Table 2.10. The high-frequency equivalent circuit of a toroidal inductor [3–10] is shown in Figure 2.13. In Figure 2.13, R represents copper losses in the winding and core losses, C is the equivalent parasitic capacitance, and L is the equivalent inductance value of the

TABLE 2.9
RF Ferrite Materials and Their Material Properties

Characteristic		T	B	G	J	K	P	Units
Initial permeability (μ_i)	15,000	10,000	5000	1500	850	125	40	
Loss factor (tan δ/μ_i)	≤7	≤7	≤15	60		150	85	×10⁻⁶
At frequency	0.01	0.01	0.1	0.1	0.1	10	10	MHz
Hysteresis factor (h/μ^2)	—	—	<2	10	6	—	—	×10⁻⁶
Saturation flux density (B_s)	370	380	450	320	280	320	215	mTesla
	3700	3800	4500	3200	2800	3200	2150	Gauss
At H maximum =	1000	1000	1000	1000	1000	2000	2000	A/m
	12.6	12.6	12.6	12.6	12.6	25	25	Oersted
Remanence (B_r)	150	140	100	150	180	160	40	mTesla
	1500	1400	1000	1500	1800	1600	400	Gauss
Coercivity (H_c)	2.4	3.2	5.6	19.9	31.8	119	278	A/m
	0.03	0.04	0.07	0.25	0.4	1.5	3.5	Oersted
Curie temperature (T_c)	≥120	≥120	≥165	≥130	≥140	≥350	≥350	°C
Temperature coefficient of μ_i (α) −40°C to +80°C (T.C.)	0.8	0.8	0.9	1.0	1.0	0.1	0.1	%/°C
Volume resistivity (p)	25	40	≥10²	≥10⁶	≥10⁵	≥10⁷	≥10⁶	Ω cm

TABLE 2.10
Composition of Ferrite Materials

Material	Initial Permeability μ_i	Composition
V	15,000	Manganese–zinc
T	10,000	Manganese–zinc
B	5000	Manganese–zinc
G	1500	Nickel–zinc
J	850	Nickel–zinc
K	125	Cobalt–nickel
P	40	Cobalt–nickel

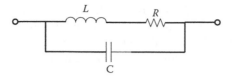

FIGURE 2.13 The equivalent circuit of toroidal core.

toroidal inductor. L is given by Equations 2.16 and 2.17. The R for the toroidal inductor is given by

$$R = R_w + R_c \tag{2.22a}$$

In Equation 2.22a, R_w is used to define the losses in the windings and R_c represents the core losses. R_c can be defined from Legg's equation to define the total core losses, including hysteresis loss, eddy current loss, and residual loss as

$$R_c = (\alpha B_m + c + ef)\mu Lf \tag{2.22b}$$

α is the hysteresis loop area constant, c is the residual loss constant, e is the eddy current coefficient, f is signal frequency, and B_m is the maximum flux density in the core. In practice, $R_c \ll R_w$ and as a result R_c can be neglected as suggested by the standard International Electrotechnical Comission (IEC):

$$R \approx R_w \tag{2.23}$$

R_w with an integer number of layers N_l is defined by Dowell [1] accurately as

$$R_w = R_{dc}A\left\{\frac{e^{2A} - e^{-2A} + 2\sin(2A)}{e^{2A} + e^{-2A} - 2\cos(2A)} + \frac{2(N_l^2 - 1)}{3}\frac{e^A - e^{-A} - 2\sin(A)}{e^A + e^{-A} + 2\cos(A)}\right\} \tag{2.24}$$

where

$$A = \left(\frac{\pi}{4}\right)^{3/4}d^{3/2}\left(\frac{\mu_{rw}\mu_0\pi f}{\rho t}\right)^{1/2} = \left(\frac{\pi}{4}\right)^{3/4}\frac{d^{3/2}}{\delta t^{1/2}} \tag{2.25}$$

and

$$R_{dc} = Nl_T\frac{4\rho}{\pi d^2} \tag{2.26}$$

In Equations 2.25 and 2.26, t is used to define the distance between the center of two adjacent turns in millimeters, N is the number of turns, l_T is the average length of a single winding in millimeters, d is the conductor diameter in millimeters, and ρ is the conductor resistivity in Ω mm ($\rho = 17.24 \times 10^{-6}$ Ω mm for copper at 20°C).

The equivalent inductance, L, is found from Equation 2.16a as

$$L = \frac{4\pi N^2 \mu_i A_{Tc}}{l_e} \text{ [nH]} \tag{2.27}$$

The equivalent parasitic capacitance of the inductor is obtained from

$$C(N) = \frac{C_{tt}}{2 + (C_{tt}/C(N-2))} + C_{tt} \tag{2.28}$$

where

$$C_{tt} = \varepsilon_0 l_t \left[\frac{\varepsilon_r \theta}{\ln(d_o/d_c)} + \cot\left(\frac{\theta}{2}\right) - \cot\left(\frac{\pi}{12}\right) \right] \tag{2.29}$$

$$\theta = \arccos\left(1 - \frac{\ln(d_o/d_c)}{\varepsilon_r} \right) \tag{2.30}$$

N is the number of turns, d_o is the wire outside diameter with coating, d_c is the wire diameter without coating, and ε_r is the permittivity of the coating used. Equation 2.28 is simplified to

$$C \cong 1.366 C_{tt} \quad \text{for } N \geq 10 \tag{2.31}$$

When the component values for the equivalent circuit in Figure 2.13 are calculated, the circuit can be reduced to equivalent series circuit shown in Figure 2.14 as previously done for air core inductors. Where the elements values of the series circuit are

$$R_s = \frac{R}{(1 - \omega^2 LC)^2 + (\omega RC)^2} \tag{2.32}$$

FIGURE 2.14 Equivalent series circuit for toroidal inductors.

and

$$L_s = \frac{(L - R^2C) - \omega^2 L^2 C}{(1 - \omega^2 LC)^2 + (\omega RC)^2} \tag{2.33}$$

The resonance frequency is then equal to

$$f_r = \frac{1}{2\pi} \sqrt{\frac{L - R^2 C}{L^2 C}} \tag{2.34}$$

Similarly, the quality factor is obtained from

$$Q = \frac{|X_s|}{R_s} \tag{2.35}$$

where $X_s = \omega L_s$.

The total power loss in the toroidal core is calculated from

$$P = 2\pi \int_{z=-h/2}^{z=h/2} \int_{\rho=id/2}^{\rho=od/2} \frac{J_\rho^2 + J_z^2}{\sigma} \rho\, d\rho\, dz \tag{2.36}$$

J_ρ and J_z are the induced current densities in ρ and z directions of the core. When the integration is performed, we obtain the total power loss at high frequencies when the skin depth is much smaller than the mean radius and height of the core as

$$P = \omega \mu_0 \mu_i I^2 N^2 \left[\frac{(h/2)\delta}{2\pi(id/2)} + \frac{(h/2)}{2\pi\sigma(od/2)} + \frac{\log(od/id)}{2\pi\sigma} \right] \tag{2.37}$$

The alternative method for core resistance calculation from Equation 2.37 is

$$R_c = \frac{P}{I^2} = \omega \mu_0 \mu_i N^2 \left[\frac{(h/2)\delta}{2\pi(id/2)} + \frac{(h/2)}{2\pi\sigma(od/2)} + \frac{\log(od/id)}{2\pi\sigma} \right] \tag{2.38}$$

Example 2.3

Design a toroidal inductor using powdered iron core material with a frequency-independent inductance value of 330 nH. The coating material used has a thickness of 0.0225 mm with permittivity 3.5. (a) Determine the number of turns, (b) Obtain the largest solid AWG wire that can be used to carry 10 A_{rms} current. (c) The maximum rms voltage drop across the inductor is required to be 30 V_{rms}. What is the maximum operational magnetic flux density? (d) What is the total core power loss? (e) Obtain the high-frequency characteristics of this inductor, identify its resonant frequency, and find its quality factor at 13.56 MHz, and quality factor versus frequency.

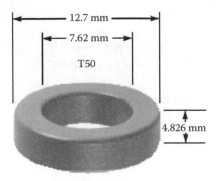

FIGURE 2.15   Geometry of T50-6 material.

SOLUTION

a. T50-2, T50-7, and T50-6 materials can be used at 13.56 MHz using Table 2.8. Since there is no specific requirement on the material type or the dimensions of the core, we choose T50-6 material to design our inductor. The geometry of the core is shown in Figure 2.15.

   The number of turns is found from

$$N = \sqrt{\frac{Ll_e}{4\pi\mu_i A_{Tc}}} = 8.87 \rightarrow 9 \text{ turns} \tag{2.39}$$

$N = 9$ turns gives $L = 339.46$ [nH].

b. The largest wire diameter is

$$d = \frac{2\pi(id)}{N + \pi} = 0.1992 \text{ [cm]} \tag{2.40}$$

which specifies 13 AWG wire as the wire to wind the core. This wire is capable of handling rms currents up to 20 A using solid 13 AWG wire from Figure 2.6. The diameter of 13 AWG wire is 1.828 mm.

c. Equation 2.20 can be used to calculate the maximum operational flux density for 30 V_{rms} voltage drop across the inductor. This gives

$$B_{op} = \frac{V_{rms} \times 10^8}{4.44fNA_{Tc}} = 45.8 \text{ [Gauss]} \tag{2.41}$$

d. The total power loss is obtained using the relation given in Equation 2.37 at the frequency of interest, 13.56 MHz, when the current flow is 10 A_{rms} as

$$P = \omega\mu_0\mu_i I^2 N^2 \left[\frac{(h/2)\delta}{2\pi(id/2)} + \frac{(h/2)}{2\pi\sigma(od/2)} + \frac{\log(od/id)}{2\pi\sigma} \right] = 4.36 \text{ [W/m}^3\text{]} \tag{2.42}$$

e. The resonant frequency is found from Equation 2.34. The values of R and C calculated from Equations 2.24 and 2.28 are

$$R = 0.06 \ [\Omega] \quad \text{and} \quad C = 5.41 \ [\text{pF}] \tag{2.43}$$

If $L = 330$ [nH] is given in the question as the desired inductance value, then the resonance frequency obtained is

$$f_r = \frac{1}{2\pi} \sqrt{\frac{L - R^2 C}{L^2 C}} = 117.45 \ [\text{MHz}] \tag{2.44}$$

The series equivalent circuit parameters are found using Equations 2.32 and 2.33 as

$$R_s = 0.06 \ [\Omega] \quad \text{and} \quad L_s = 339.6 \ [\text{nH}] \tag{2.45}$$

Then, the quality factor at 13.56 MHz is

$$Q = \frac{|X_s|}{R_s} = 70.36 \tag{2.46}$$

The high-frequency response and quality factor versus frequency are given in Figures 2.16 and 2.17, respectively.

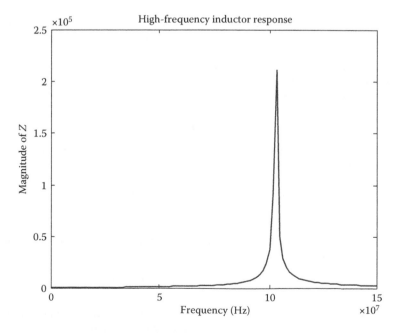

FIGURE 2.16 High-frequency characteristics response of the toroidal inductor.

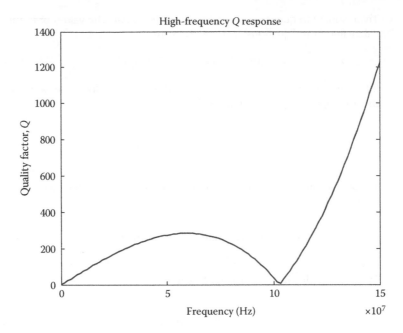

FIGURE 2.17   Quality factor of the toroidal inductor versus frequency for Example 3.

Design Example : Stacked Toroidal Inductor Design

Design a toroidal inductor that is formed by stacking two T106-6 powdered iron cores with 21 turns using 16 AWG wire at 2 MHz. (a) What are the theoretical and measured inductance values? (b) Compare the calculated and measured quality factor. Plot the calculated quality factor versus frequency. (c) Calculate the resonant frequency and obtain high-frequency characteristics. (d) Calculate the wire length used to wind the core.

SOLUTION

a. The calculated frequency-independent inductance value is found from Equation 2.16a as

$$L = \frac{4\pi N^2 \mu_i A_{Tc}}{l_e} = \frac{4\pi (21)^2 (8.5)(2 \times 0.6908)}{6.3026} = 10.325 \ [\mu H]$$

The series equivalent circuit parameters are found using Equations 2.32 and 2.33 at 2 MHz as

$$R_s = 1.1951 \ [\Omega] \quad \text{and} \quad L_s = 10.6 \ [\mu H]$$

The inductor is assembled as shown in Figure 2.18 and measured with HP impedance analyzer 4191A and found to be $L = 10.38 \ [\mu H]$.

b. The calculated quality factor at 2 MHz is

(a)   (b)

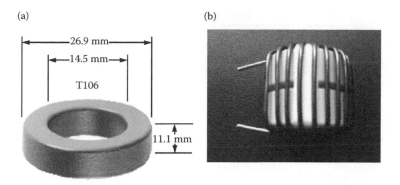

FIGURE 2.18 (a) Physical dimensions for T106 core. (b) The assembled inductor with T106-6 stacked cores.

$$Q = \frac{|X_s|}{R_s} = 111.46$$

The assembled inductor quality factor is measured with HP impedance analyzer 4191A and is found to be $Q = 101$. The quality factor versus frequency is plotted and illustrated in Figure 2.19.

c. The value of R, C, and L are calculated as

$$R = 1.1342 \ [\Omega], \quad C = 15.84 \ [pF], \quad \text{and} \quad L = 10.325 \ [\mu H]$$

The frequency characteristics are illustrated in Figure 2.20. The resonance frequency is obtained from

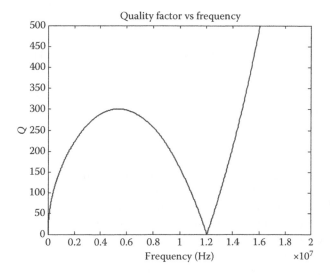

FIGURE 2.19 Quality factor of the toroidal inductor versus frequency.

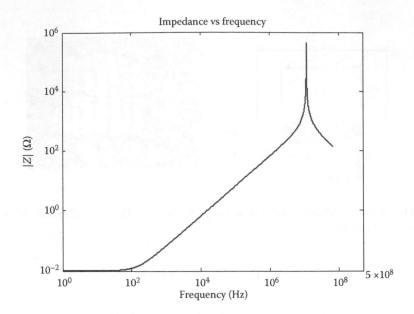

FIGURE 2.20   High-frequency characteristics of T106-6 stacked cores.

$$f_r = \frac{1}{2\pi}\sqrt{\frac{L - R^2C}{L^2C}} = 12.46 \text{ [MHz]}$$

d. The wire length is calculated using Equation 2.19 as

$$l_w = 2\pi\sqrt{[N(2r + od - id)]^2 + (od + id)^2} = 149.5 \text{ [cm]}$$

2.4   TOROIDAL INDUCTOR DESIGN AND CHARACTERIZATION PROGRAM

MATLAB GUI has been developed to design toroidal inductors and provide several critical design parameters and high-frequency characteristics. The program produces the following outputs as design parameters:

- L_s, series equivalent inductor value that the measurement device displays
- R_s, series equivalent resistance value that the measurement device displays
- Q, quality factor value that the measurement device displays
- R_{ac}, AC resistance of the toroidal configuration
- R_w, winding resistance of the toroidal inductor
- R_c, core resistance of the toroidal inductor
- L, frequency-independent inductor value
- C, capacitance of the toroidal configuration
- F_r, resonant frequency of the toroidal configuration

- l_w, wire length required for the number of turns desired
- l_e, magnetic path length of the core in centimeter
- A_e, cross-sectional area of the core in centimeter^2

The high-frequency characteristics of the toroidal inductor are displayed via several characteristic curves. The input parameters to design the toroidal inductor and obtain the frequency characteristics are

- N, number of turns
- od, outside diameter of the toroidal core in centimeter
- id, inside diameter of the toroidal core in centimeter
- h, thickness of the toroidal core in centimeter
- n, number of stacked cores
- nui, initial permeability
- d, wire diameter without coating in centimeter
- s, coating thickness of the wire in millimeter
- epsr, coating material permittivity
- F_o, frequency of interest

The screen capture of the program to accomplish the complete design of the toroid is displayed in Figure 2.21. The program is developed using the formulas given between Equations 2.16 and 2.46.

The results that are obtained from the program are tested experimentally by assembling the toroidal inductors using T106-6 material as shown in Figure 2.22. Four different configurations at three different frequencies have been measured using HP 4191A impedance analyzer. The measured results are compared with the analytical results and given in Figures 2.23 through 2.25.

As seen from Figures 2.23 through 2.25, as the number of turns and frequency increase, the error between the measured values and calculated values increases for inductance, quality factor, and series equivalent resistance. HP 4191A impedance

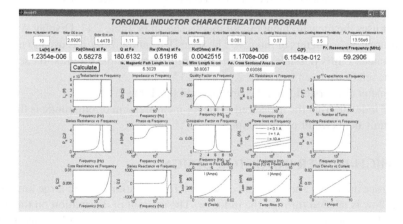

FIGURE 2.21 Toroid characterization program.

FIGURE 2.22   Assembled toroidal inductors.

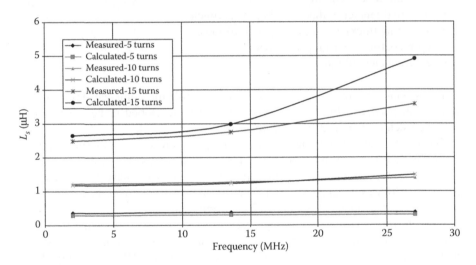

FIGURE 2.23   Comparison of measured and calculated inductance values.

analyzer uses the measurement circuit modes shown in Table 2.11. Display A is used to measure the parameters of interest. The series RL sample circuit is used to measure the parameters. It is possible to convert the equivalent sample circuits to each other using the relations given in Tables 2.12 and 2.13.

Parameter conversion formulas are given in the Table 2.12.

The * indicates that the measured value for measurement circuit mode is the same as the equivalent circuit of the sample.

2.5   HIGH-POWER INDUCTOR DESIGN

High-power inductors are commonly used in RF applications where large amount of current flow is needed through the inductor. The amount of dissipation on the windings can be reduced significantly by using copper strip with rectangular cross section for winding. Increased cross-sectional area by using copper strip helps reduce the

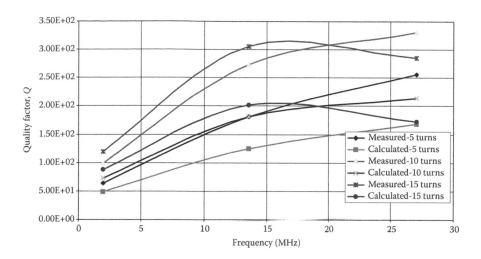

FIGURE 2.24 Comparison of measured and calculated quality factor values.

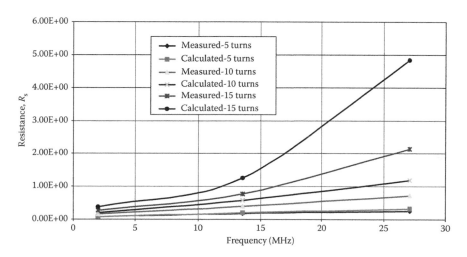

FIGURE 2.25 Comparison of measured and calculated resistance values.

dissipation and associated heat in the windings. The *dc* resistance with the copper is now calculated from

$$R_{dc} = \frac{l_{Tr}}{\sigma A} = \frac{l_{Tr}}{\sigma(W \times T)} \tag{2.47}$$

In Equation 2.47, l_{Tr} is the length of the copper strip, W and T represent the width and thickness of the copper strip. Equation 2.47 can be substituted into Equation 2.24 for the calculation of the winding resistance or *ac* resistance. The procedure described in Section 2.3 remains the same for the calculation of critical parameters

TABLE 2.11

Measurement Circuit Mode

Display A	Display B	Circuit Mode
R	X	
G	B	
L	G	
	R, D, Q	
C	R	
	G, D, Q	

TABLE 2.12

Parameter Conversion Table

Sample	$|Z|$	$|Y|$	R	G	L	C
C R	$\sqrt{\dfrac{1}{\omega^2 C^2} + R^2}$	$\dfrac{\omega C}{\sqrt{1 + \omega^2 C^2 R^2}}$	R	$\dfrac{\omega^2 C^2 R}{1 + \omega^2 C^2 R^2}$	—	C^*
C G	$\dfrac{1}{\sqrt{\omega^2 C^2 + G^2}}$	$\sqrt{\omega^2 C^2 + G^2}$	$\dfrac{G}{G^2 + \omega^2 C^2}$	G	—	C^*
L R	$\sqrt{\omega^2 L^2 + R^2}$	$\dfrac{1}{\sqrt{\omega^2 L^2 + R^2}}$	R	$\dfrac{R}{R^2 + \omega^2 L^2}$	L^*	—
L G	$\dfrac{\omega L}{\sqrt{1 + \omega^2 L^2 G^2}}$	$\sqrt{\dfrac{1}{\omega^2 L^2} + G^2}$	$\dfrac{\omega^2 L^2 G}{1 + \omega^2 L^2 G^2}$	G	L^*	—

(Display A Parameters)

such as resonant frequency, quality factor, and so on for the inductor. The amount of flux in the core can be reduced by establishing a symmetric core structure with the use of identical toroidal cores as shown in Figure 2.26. The symmetrical core structure resembles the binocular core in shape and carries the advantages of toroidal cores in operation with an increased core cross-sectional area that reduces the amount of flux. The calculation of the design parameters can be done using the toroidal software GUI that is developed with the physical dimension given in Figure 2.27.

The high-power inductors are assembled using 4xT106-6 iron powdered cores in two symmetrical stack configurations as shown in Figure 2.28. Four-turn and

TABLE 2.13
Dissipation Factor Equations

	Circuit Mode	Dissipation Factor	Conversion to Other Modes
C	C_p parallel R_p	$D = \dfrac{1}{2\pi f C_p R_p} = \dfrac{1}{Q}$	$C_s = (1 + D^2)C_p$ $R_s = \dfrac{D^2}{1 + D^2} R_p$
	$C_s\ R_s$ series	$D = 2\pi f C_s R_s = \dfrac{1}{Q}$	$C_p = \dfrac{1}{1 + D^2} C_s$ $R_p = \dfrac{1 + D^2}{D^2} R_s$
L	L_p parallel R_p	$D = \dfrac{2\pi f L_p}{R_p} = \dfrac{1}{Q}$	$L_s = \dfrac{1}{1 + D^2} L_p$ $R_s = \dfrac{D^2}{1 + D^2} R_p$
	$L_s\ R_s$ series	$D = \dfrac{R_s}{2\pi f L_s} = \dfrac{1}{Q}$	$L_p = (1 + D^2)L_s$ $R_p = \dfrac{1 + D^2}{D^2} R_s$

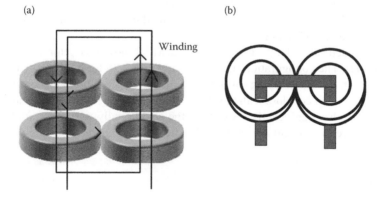

(a)   (b)

Winding

FIGURE 2.26 Construction of high-power inductor with toroidal cores (a) direction of winding (b) winding with copper strip.

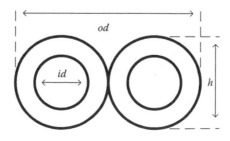

od

id

h

FIGURE 2.27 Physical dimensions of high-power toroidal inductor.

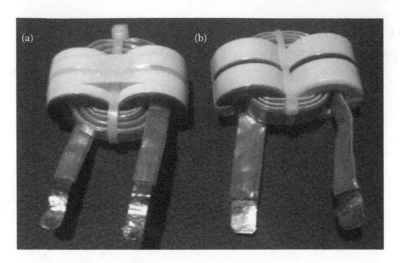

FIGURE 2.28　Assembled (a) five-turn (b) four-turn high-power inductor with toroidal cores.

five-turn high-power inductor using copper strip with 0.85 and 0.7 cm width, respectively and 0.06 cm thickness for winding are shown in Figure 2.28.

The multiaperture core inductors are measured with HP 4191A impedance analyzer. Their inductance values at 2 MHz are measured as 1118 [nH] and 1755 [nH], respectively. The analytical calculation is performed using the software and is illustrated in Figure 2.29 for a five-turn inductor.

The calculated values for the four- and five-turn high-power inductors are 988 [nH] and 1561 [nH]. The configuration illustrated in Figure 2.28 for high-power inductor gives very high current-carrying capacity and increased saturation flux density. As a result, the amount of core loss with this type of configuration reduces drastically. The copper loss is also significantly lower than the traditionally wound toroidal cores. This can be seen in the curves obtained by the toroidal characterization program.

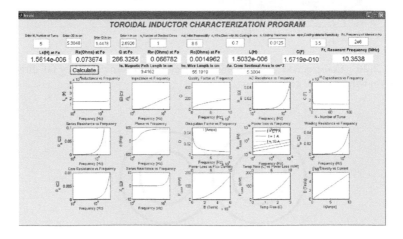

FIGURE 2.29　Five-turn high-power inductor design with toroidal characterization program.

The quality factor at the frequency of interest is also increased and contributes to the thermal characteristics of the core during operation. However, one disadvantage of this configuration is reduced resonance frequency due to significantly increased stray capacitance shown in Figure 2.13.

2.6 PLANAR INDUCTOR DESIGN AND EXAMPLES

Spiral-type planar inductors are widely used in the design of power amplifiers (PAs), oscillators, microwave switches, combiners, splitters, and so on for industrial, scientific, and medical (ISM) applications at the HF range. HF range is a common frequency range that is used for ISM applications. Spiral inductors, when used in ISM applications, must be designed to handle power in the range of several kilowatts. They should demonstrate good thermal characteristics, sustained inductance value, and low loss under such high power because any change in the component values in RF system affects the performance and can cause catastrophic failures. This can be prevented by the application of an accurate design method for the material that is used as a substrate.

In this section, practical and accurate analytical design method and algorithm for low-profile microstrip spiral inductors are given using the simplified lumped-element equivalent model. The method and algorithm are used to obtain the physical dimensions and resonant frequency of the spiral inductors for the desired inductance values. The physical dimensions are then used to simulate the spiral inductors by planar electromagnetic simulator, Sonnet, to validate the design parameters. Spiral inductors are then implemented on 100-mil-thick alumina (Al_2O_3) substrate using the implementation method proposed for the required level of power dissipation. The network analyzer, HP8753ES, is used to measure the inductance values and resonant frequencies of the spiral inductors.

2.6.1 ANALYTICAL DESIGN METHOD FOR SPIRAL INDUCTOR AT HF RANGE

The inductance value of the spiral inductors at the HF range can be determined using the quasistatic method proposed by Greenhouse [14] with a good level of accuracy.

The method proposed by Greenhouse takes into account the self-coupling and mutual coupling between each trace. The layout of the two conductors that is used in the inductance calculation is illustrated in Figure 2.30a. GMD is the geometric mean distance between two conductors and AMD represents the arithmetic mean distance between two conductors. The total inductance of the configuration of the spiral inductor is

$$L_T = L_0 + \Sigma M \tag{2.48a}$$

where L_T is the total inductance, L_0 is the sum of the self-inductances, and ΣM is the sum of the total mutual inductances. The application of the formulation given by Dowell [1] can be demonstrated for the spiral inductor illustrated in Figure 2.1b and 2.1c as

$$L_T = L_1 + L_2 + L_3 + L_4 + L_5 - 2(M_{13} + M_{24} + M_{35}) + 2M_{15} \tag{2.48b}$$

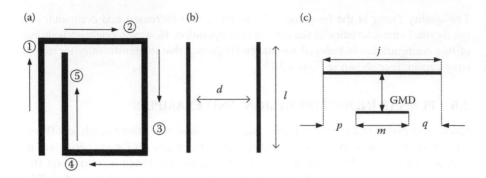

FIGURE 2.30 (a) Rectangular spiral inductor. (b) Layout of current filaments. (c) Two-parallel filament geometry.

The general relations that can be used in the algorithm for the spiral inductance calculation then become

$$L_i = 0.0002l_i \left[\ln\left(2\frac{l_i}{\text{GMD}} \right) - 1.25 + \frac{\text{AMD}}{l_i} + \frac{\mu}{4}T \right] \tag{2.49}$$

$$M_{ij} = 0.0002l_iQ_i \tag{2.50}$$

$$\ln\left(\text{GMD}_i\right) = \ln(d) - \frac{1}{12(d/w)^2} - \frac{1}{60(d/w)^4} - \frac{1}{168(d/w)^6} - \frac{1}{360(d/w)^8} - \cdots \tag{2.51}$$

$$Q_i = \ln\left[\frac{l_i}{\text{GMD}_i} + \left(1 + \left(\frac{l_i}{\text{GMD}_i} \right)^2 \right)^{0.5} \right] - \left(1 + \left(\frac{\text{GMD}_i}{l_i} \right)^2 \right)^{0.5} + \frac{\text{GMD}_i}{l_i} \tag{2.52}$$

$$\text{AMD} = w + t \tag{2.53}$$

where L_i is the self-inductance of the segment i, M_{ij} is the mutual inductance between segments i and j, l_i is the length of the segment i, μ is the permeability of the conductor, T is the frequency correction factor, d is the distance between the conductor filaments, w is the width of the conductor, t is the thickness of the conductor, Q_i is the mutual inductance parameter of segment i, GMD_i is the geometric distance of segment i, and AMD is the arithmetic mean distance.

The two-port lumped-element equivalent circuit for a spiral inductor at the HF range is represented by a π-network shown in Figure 2.31.

C_T is the lumped-element equivalent circuit capacitance that has the effect of odd mode, even mode, and interline coupling capacitances between coupled lines of the spiral inductor. The spiral inductor is separated into segments for the overall capacitance calculation. The lines that are coupled and used for odd mode, even mode, and

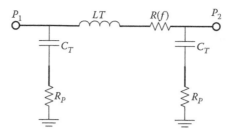

FIGURE 2.31 Two-port lumped-element equivalent circuit for the spiral inductor.

interline coupling capacitances calculation are circled in Figure 2.32. One half of the spiral inductor is used for the calculation of C_T due to symmetry.

The capacitance calculation for additional segments that are out of the symmetric structure of the spiral inductor has to be performed and added to the overall capacitance. The capacitance between spiral segments is found from the equivalent circuit shown in Figure 2.33.

C_e is the even mode capacitance, C_o is the odd mode capacitance, and C_c is the interline coupling capacitance in Figure 2.33. The interline coupling capacitance is found from

$$C_c = \frac{1}{2}(C_o - C_e)\,[\text{F/m}] \qquad (2.54)$$

After the calculation of the coupling capacitances between the coupled spiral segments, the equivalent capacitance, C_{eq}, including the effect of all segments, is found from the network theory. The spiral inductor capacitance C_T is then found using the following relation:

$$C_T = 2C_{eq} + C_{o(outside)}\,[\text{F/m}] \qquad (2.55)$$

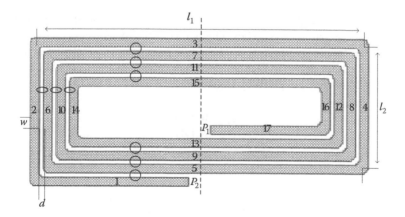

FIGURE 2.32 Segmentation of the spiral inductor for capacitance calculation.

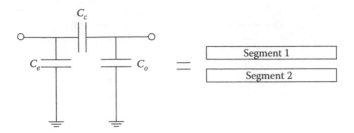

FIGURE 2.33   Equivalent circuit used for capacitance calculation between spiral segments.

where $C_{o(outside)}$ is the odd mode capacitance of the outside turn. All the capacitances are calculated per unit length. The series resistance $R(f)$ is found by assuming transverse electromagnetic (TEM) wave propagation on the lines of the spiral inductor. The resistance, $R(f)$, can then be calculated as

$$R(f) = \frac{2}{w}\sqrt{\frac{\pi f \mu_c}{\sigma}} \quad [\Omega/m] \tag{2.56}$$

μ_c and σ are the permeability and conductivity of the conductor, respectively, that is used as a trace for the spiral inductor. R_p is the substrate resistance and is calculated using

$$R_p = \frac{\rho l}{wh} \quad [\Omega/m] \tag{2.57}$$

where w is the width of the trace, h is the thickness of the substrate, and ρ is given in Equation 2.26 for Al_2O_3. One-port measurement network is obtained from the circuit shown in Figure 2.31 by grounding the second port and is illustrated in Figure 2.34.

The quality factor, Q, and the resonant frequency, f_r, of the spiral inductor are important design parameters. They characterize the performance of the spiral inductor at the frequency of interest. These two parameters are calculated using the measurement circuit shown in Figure 2.34. The expressions for the quality factor and the resonant frequency are given by Equations 2.58 and 2.59 as

$$Q = \frac{\text{Im}(Z)}{R_e(Z)} \tag{2.58}$$

and

$$f_r = \frac{1}{2\pi\sqrt{L_T C_T}} \left[\frac{1 - R^2(C_T/L_T)}{1 - R_p^2(C_T/L_T)} \right]^{0.5} \tag{2.59}$$

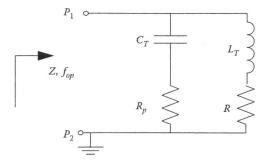

FIGURE 2.34 Spiral inductor measurement circuit.

2.6.2 MATERIAL DESIGN

The material that is used as a substrate for the spiral inductor should be able to handle the power dissipation during the operation. The substrate material at the frequency of the operation should present low loss and have high thermal conductivity. One of the commonly used materials exhibiting these features is Al_2O_3. Thermal resistance of Al_2O_3 is found from

$$\theta_D = \frac{h}{kA} \ [°C/W] \qquad (2.60)$$

θ_D is the thermal resistance in °C/W, h is the thickness in inches, k is the thermal conductivity of the substrate in watts/(in °C), and A is the area in in². Since $k = 0.6$ [watts/(in °C)] is a known parameter, the thickness and the area of the substrate can be selected to meet with the required power dissipation. This can be better visualized on an arbitrary-shaped material specimen shown in Figure 2.35. Figure 2.35 shows a layer of a thermally conductive material that has a constant cross-sectional area A, thickness h, and thermal conductivity k. If the surface S_1 is maintained at a temperature T_1 and surface S_2 is maintained at a higher temperature T_2, then the dissipated power in the layer is defined as

$$P_D = \frac{T_2 - T_1}{\theta_T} \ [W] \qquad (2.61)$$

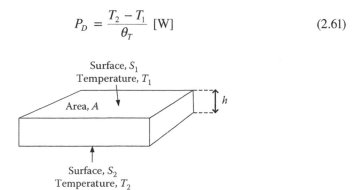

FIGURE 2.35 Material specimen for thermal calculation.

The purity of Al_2O_3 is given to be 96%. The metalization on the back and front of the alumina substrate is the composition of copper, nickel, and gold. The alumina spiral inductor is usually placed on a cold plate with an application of a thermal compound, which has a thermal resistance around $\theta_C = 0.01°C/W$. The power-handling capacity of Al_2O_3 versus temperature, including cold plate and thermal compound, is illustrated in Figure 2.36. The operating substrate temperature of Al_2O_3 can go up to 300°C. The calculated worst-case temperature during operation should be below the safe operating range of the material. As a result, the substrate properties, including physical dimensions such as thickness, surface area, and thermal characteristics can be determined for a reliable operation in RF applications using the method described above with the application of Equations 2.60 and 2.61.

2.6.2.1 Design Examples: Spiral Inductor Design with Alumina Substrate

On the basis of the formulation introduced in Section 2.6.1, several spiral inductors are designed, simulated, and built using Al_2O_3 as a substrate with the method given in Section 2.6.2. Constructed microstrip spiral inductors are measured using HP 8753ES network analyzer.

Using the formulation given in Section 2.6.1, five spiral inductors are designed with different number of turns, trace width, spacing, conductor length, and configurations. The substrate material is selected to be Al_2O_3 with relative permittivity constant 9.8 and thickness 0.254 cm. The operational frequency for the inductors is 13.56 MHz, which is a common frequency for ISM applications. Table 2.14

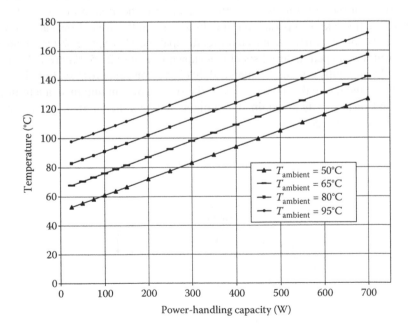

FIGURE 2.36 Power-handling capacity for 0.1-in-thick Al_2O_3 including cold plate and thermal compound.

TABLE 2.14

Physical Dimensions of Spiral Inductors

	Trace Width (cm)	Spacing (cm)	L_1 (cm)	L_2 (cm)
Inductor 1	0.2032	0.1016	7.7	2.95
Inductor 2	0.2032	0.0508	7.8	3
Inductor 3	0.1524	0.0762	7.7	1.4
Inductor 4	0.1524	0.0762	7.7	1.9
Inductor 5	0.2032	0.3048	9.65	1.7

TABLE 2.15

Analytical Results for Spiral Inductors

	Number of Turns	Inductance (nH)	Resonant Frequency f_r (MHz)
Inductor 1	4	516.41	38.96
Inductor 2	5	753.73	30.49
Inductor 3	2.5	238.41	69.22
Inductor 4	3.5	405.08	53.29
Inductor 5	1.5	132.11	81.62

illustrates the dimensions of the spiral inductors that are designed. An algorithm is developed using Mathcad to calculate the lumped-element equivalent circuit element values and plot them on the Smith chart. Table 2.15 shows the analytical results for the calculated inductance value, and the resonant frequency for each inductor. The output of the developed program plotting the equivalent impedance on the Smith chart for inductor 1 is shown in Figure 2.37. During the operation of any RF system, spiral inductors are desired to have resonant frequencies approximately 3 times higher than the operational frequency of the system as a rule of thumb. The quality factor is needed to be high to avoid any thermal problems due to conductor and dielectric losses. On the basis of the analytical results, these two requirements are met for all the inductors except the second inductor for the resonant frequency as shown in Table 2.15.

2.6.2.2 Simulation Results

Rectangular spiral inductors whose physical dimensions are given in Table 2.14 are simulated using method of moment (MoM)-based planar electromagnetic simulator, Sonnet. The simulation results will be used to validate the analytical results and define the final configuration for the inductors that will be built. The layouts of the simulated spiral inductors are illustrated in Figure 2.38. Inductors 3 and 4 are implemented on the same board. Inductor 5 is implemented as a symmetrical inductor also on a single board. Inductors 1 and 2 are implemented on separate boards. Different configurations of the spirals inductors that are illustrated in Figure 2.38 demonstrate how to implement a spiral inductor based on the application to have

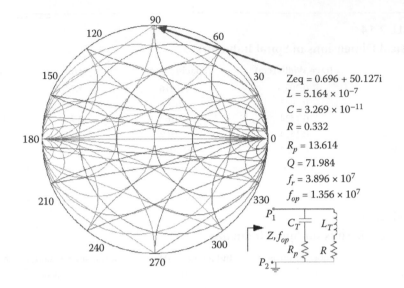

Zeq = 0.696 + 50.127i
$L = 5.164 \times 10^{-7}$
$C = 3.269 \times 10^{-11}$
$R = 0.332$
$R_p = 13.614$
$Q = 71.984$
$f_r = 3.896 \times 10^7$
$f_{op} = 1.356 \times 10^7$

FIGURE 2.37 Spiral inductor lumped-element calculation program.

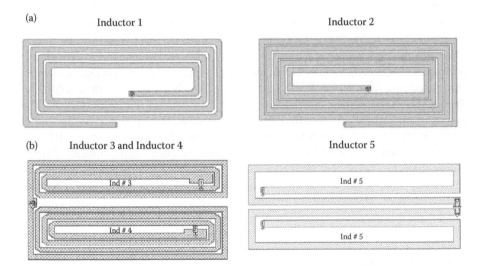

FIGURE 2.38 Layout of the spiral inductors that are simulated. (a) Inductors 1 and 2. (b) Inductors 3, 4, and 5.

optimum performance and results. Inductors 1 and 2 are used as components for the filter in RF PA system whereas inductors 3, 4, and 5 are used as resonating inductors for RF PA. The simulation results for the five spiral inductors are given in Table 2.16.

The error between the simulation results and analytical results is given in Table 2.17. The maximum error in the inductance calculation is found to be around 3%.

TABLE 2.16

Simulation Results for Spiral Inductors

	Inductance (nH)	Resonant Frequency f_r (MHz)
Inductor 1	529.76	40.8
Inductor 2	741.41	33
Inductor 3	244.6	77.2
Inductor 4	392.97	54.25
Inductor 5	132.78	92.5

TABLE 2.17

Error between Simulation Results and Analytical Results

	Error in Inductance (%)	Error in Frequency (%)
Inductor 1	2.52	4.51
Inductor 2	1.66	7.61
Inductor 3	2.53	10.34
Inductor 4	3.08	1.77
Inductor 5	0.50	11.76

The error increases to 11% for resonant frequency calculation. Note that although our lumped-element model and formulation for analytical calculation is one of the most involved model to date using most of the effects such as coupling capacitances, mutual inductances, substrate resistance, series resistance, skin effect, and so on, it is still an approximate model with a certain level of inaccuracy.

One of the biggest advantages of the simulation programs is the visualization of the current density distribution of the spiral inductor. The simulation result for the first inductor illustrating the current density distribution is given in Figure 2.39. This is a very important information in practice, which can be used to adjust the spacing between the traces to prevent potential arcs during operation by distributing the voltage drop between the turns of the spiral at the desired level.

2.6.2.3 Experimental Results

The spiral inductors that are designed and simulated are built and measured using the network analyzer HP 8753ES. Figure 2.40 illustrates the spiral inductors that are manufactured. The measurement results are given in Table 2.18, whereas the error between the measured, simulation, and analytical results are given in Table 2.19. It has been observed that simulation results generally gave better error for inductance calculation and resonant frequency for the spiral inductors.

However, it is worth noting that the error for the inductance value and resonant frequency were close between analytical, simulation, and experimental results. This confirms the accuracy of the proposed analytical method and developed algorithm to design spiral inductors at the HF range for ISM applications.

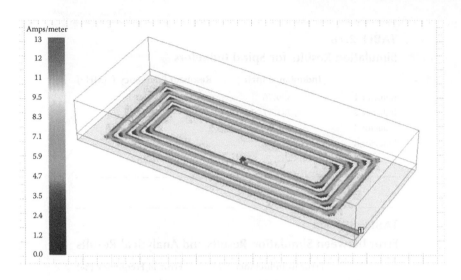

FIGURE 2.39 Simulated spiral inductor showing the current density.

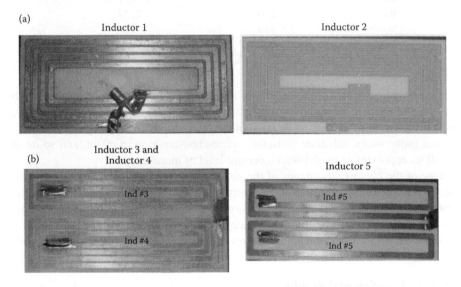

FIGURE 2.40 The manufactured spiral inductors. (a) Inductors 1 and 2. (b) Inductors 3, 4, and 5.

2.7 MULTILAYER AND COMPOSITE INDUCTOR DESIGN AND EXAMPLES

The advancement of technology dictates the use of low-profile components with high-power capacity. Planar spiral inductors as discussed in Section 2.6 have low-profile and good thermal characteristics when a ceramic-type substrate such as alumina is used. The performance of the planar inductor discussed in Section 2.6 can be further enhanced by forming a composite inductor structure with multilayer

TABLE 2.18

Measurement Results

	Inductance (nH)	Resonant Frequency f_r (MHz)
Inductor 1	529.04	41.22
Inductor 2	762.13	32.94
Inductor 3	257.04	73.9
Inductor 4	413.5	52.74
Inductor 5	137.88	88.56

TABLE 2.19

Error between Measured, Simulation, and Analytical Results

	Measurement Inductance Value (nH)	Analytical Inductance Value (nH)	Simulation Inductance Value (nH)	Inductance Error (%) Measurement versus Analytical	Inductance Error (%) Measurement versus Simulation
Inductor 1	529.04	516.41	529.76	2.45	0.14
Inductor 2	762.13	753.73	741.41	1.11	2.79
Inductor 3	257.04	238.41	244.6	7.81	5.09
Inductor 4	413.5	405.08	392.97	2.08	5.22
Inductor 5	137.88	132.11	132.78	4.37	3.84

	Measurement Resonant Frequency (MHz)	Analytical Resonant Frequency (MHz)	Simulation Resonant Frequency (MHz)	Frequency Error (%) Measurement versus Analytical	Frequency Error (%) Measurement versus Simulation
Inductor 1	41.22	38.96	40.8	5.80	1.03
Inductor 2	32.94	30.49	33	8.04	0.18
Inductor 3	73.9	69.22	77.2	6.76	4.27
Inductor 4	52.74	53.29	54.25	1.03	2.78
Inductor 5	88.56	81.62	92.5	8.50	4.26

configuration using magnetic overlay material. When the conductor carries electric current, it induces the magnetic field around it as shown in Figure 2.41.

The magnetic field that is induced is characterized by magnetic field intensity, $\bar{H}$. This relation can be mathematically expressed as

$$L = \frac{N\Psi}{I} = \frac{N\int \bar{B} \cdot d\bar{s}}{I} \qquad (2.62)$$

Now, consider one of the single rectangular loop shown in Figure 2.42.

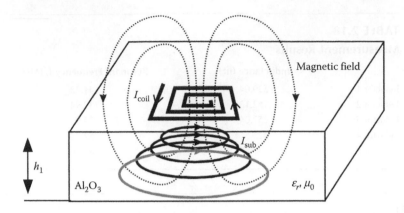

FIGURE 2.41   Induced magnetic field for spiral inductors.

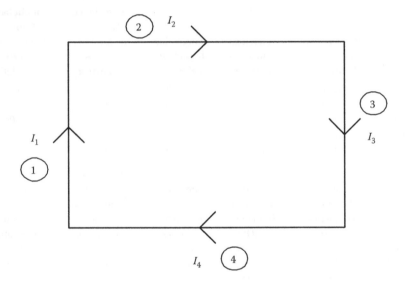

FIGURE 2.42   Single rectangular loop.

The inductance for the single loop is then equal to

$$L = \frac{\int \bar{B} \cdot d\bar{s}}{I} \tag{2.63}$$

Using Stoke's theorem, the inductance of the loop can be written in terms of the magnetic potential as

$$L = \frac{\int_S \bar{B} \cdot d\bar{s}}{I} = \frac{\oint_C \bar{A} \cdot d\bar{l}}{I} \tag{2.64}$$

where $\bar{B} = \nabla \times \bar{A}$. This integral is expanded as

$$L = \frac{\oint_C \bar{A} \cdot d\bar{l}}{I} = \frac{1}{I}\left[\oint_{C_1} \bar{A} \cdot d\bar{l_1} + \oint_{C_2} \bar{A} \cdot d\bar{l_2} + \oint_{C_3} \bar{A} \cdot d\bar{l_3} + \oint_{C_4} \bar{A} \cdot d\bar{l_4}\right] = L_1 + L_2 + L_3 + L_4$$

(2.65)

L_i, which is expressed above is called as self-inductance. The net inductance for each segment is expressed as the sum of self-inductance and mutual inductance and can be written as

$$L_i = \sum_{j=1}^{k} \pm L_{p_{ij}}$$

(2.66)

where k represents the total number of segments for the spiral inductor. When $i = j$, Equation 2.66 leads to the self-inductance and when $i \neq j$, it leads to mutual inductance of each segment. When this is repeated for all segments, this leads to the total inductance of the spiral inductor given in Equation 2.67 as

$$L_T = \sum_{i=1}^{k} L_i = \sum_{i=1}^{k}\sum_{j=1}^{k} \pm L_{p_{ij}} = L_0 + \Sigma M$$

(2.67)

The mutual inductance in Equation 2.66 is calculated using

$$L_{p_{ij}} = \frac{\oint_{C_i} \bar{A}_{ij} \cdot d\bar{l_i}}{I_j}$$

(2.68)

Equation 2.68 represents the ratio of the magnetic flux penetrating the surface between segment i and infinity and current I_j that produces that flux.

Now, consider adding another layer on top of the spiral conductor, which is laid on the ceramic substrate with a finite thickness to have the new configuration in Figure 2.43.

The addition of the magnetic substrate increases the flux density in the upper portion of the configuration where flux lines pass through the magnetic material instead of free space. As a consequence, this results in an increase of the overall inductance value. It also improves the quality factor of the spiral inductor.

Figure 2.44 illustrates the typical design example of the multilayer composite spiral inductor using Al_2O_3 as the main substrate and ferrite as the top layer. The thickness of Al_2O_3, h_1, is set to be 100 mil whereas the thickness of the top ferrite layer, h_2, is taken to be 300 mil. Ferrite materials with different permeabilities ranging from 20 to 120 are used to see the enhancement in the inductance and quality factor.

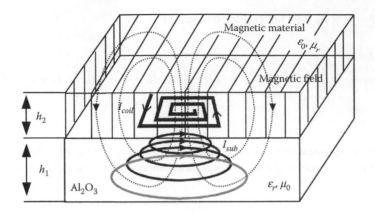

FIGURE 2.43 Spiral inductor with magnetic material.

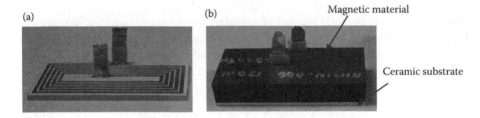

FIGURE 2.44 (a) Spiral inductor on Al_2O_3. (b) Composite spiral inductor on Al_2O_3 with ferrite top layer.

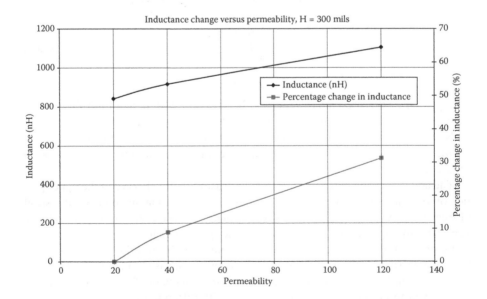

FIGURE 2.45 Change in the inductance value when ferrite magnetic layer is used.

The inductance value of the spiral inductor without ferrite material is measured to be 637.19 [nH]. The data in Figure 2.45 show the change in the inductance value versus the permeability of ferrite.

REFERENCES

1. P.L. Dowell. Effects of eddy current in transformer windings, *Proceedings of IEE*, 113(8), 1287–1394, 1966.
2. E.C. Snelling. *Soft Ferrites, Properties and Applications*, ILIFFE Books, London, UK, 1969.
3. Q. Yu, T.W. Holmes, and K. Naishadham. RF equivalent circuit modeling of ferrite core inductors and characterization of core materials, *IEEE Transactions on Electromagnetic Compatibility*, 44(1), 258–262, 2002.
4. L. Dalessandro, W.G.H. Odendaal, and J.W. Kolar. HF characterization and nonlinear modeling of a gapped toroidal and magnetic structure, *IEEE Transactions on Power Electronics*, 21(5), 1167–1175, 2006.
5. Y.S. Yuan. Modeling of stray capacitance for inductor, *Journal of East China Jiaotong University*, 23(5), 90–93,101, 2006.
6. M.P. Perry. Multiple layer series connected winding design for minimum losses, *IEEE Transactions on Power Apparatus and Systems*, PAS-98, 116–123, 1979.
7. R.W. Rhea. A multimode high-frequency inductor model, *Applied Microwave & Wireless*, 9(6), 70–80, 1997.
8. J.P. Vandelac and P.D. Ziogas. A novel approach for minimizing high frequency transformer copper losses, *IEEE Transactions on Power Electronics*, 3, 266–276, 1988.
9. M. Bartoli, A. Reatti, and M.K. Kazimierczuk. Modeling iron-powder inductors at high frequencies, *Proceedings of the 1994 IEEE-IAS Annual Meeting*, 2, 1225–1232, 1994.
10. A. Massarini and M.K. Kazimierczuk. Self-capacitance of inductors, *IEEE Transactions on Power Electronics*, 12, 671–676, 1997.
11. Micrometals, http://www.micrometals.com.
12. Ferronics Inc., http://www.ferronics.com.
13. National Magnetics, http://www.magneticsgroup.com.

3 MF-UHF Transformer Design Techniques

3.1 INTRODUCTION

A transformer is a device that converts one level of voltage to another level at the same radio frequency (RF). It consists of one or more coil(s) of wire wrapped around a common magnetic core. These coils are usually not connected together electrically. However, they are connected through the common magnetic flux confined to the core. Assuming that the transformer has at least two windings, one of them (primary) is connected to a source of RF power and the other (secondary) is connected to the RF loads. The typical toroidal transformer is shown in Figure 3.1.

The relationship between the voltage applied to the primary winding $v_p(t)$ and the voltage produced on the secondary winding $v_s(t)$ can be established as follows. Assume that $v_p(t)$ is applied to the primary winding of the toroid in Figure 3.1. The average flux in the primary winding can be written as

$$\bar{\phi} = \frac{1}{N_p} \int v_p(t)\, dt \tag{3.1}$$

A portion of the flux produced in the primary coil passes through the secondary coil as mutual flux and the rest is lost as leakage flux, which can be shown using the following relation:

$$\bar{\phi}_p = \phi_m + \phi_{Lp} \tag{3.2}$$

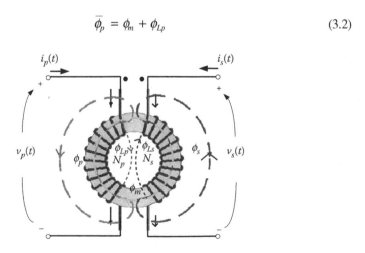

FIGURE 3.1 Toroidal transformer.

In the secondary coil, a similar relation holds as

$$\bar{\phi}_s = \phi_m + \phi_{Ls} \tag{3.3}$$

Using Faraday's law, we can write the voltage on the primary winding as

$$v_p(t) = N_p \frac{d\bar{\phi}_p}{dt} = N_p \frac{d\phi_m}{dt} + N_p \frac{d\phi_{Lp}}{dt} = v_1(t) + v_{1L}(t) \tag{3.4}$$

Then, the secondary voltage can be expressed as

$$v_s(t) = N_s \frac{d\bar{\phi}_s}{dt} = N_s \frac{d\phi_m}{dt} + N_s \frac{d\phi_{Ls}}{dt} = v_2(t) + v_{2L}(t) \tag{3.5}$$

The voltage on the primary winding and secondary winding due to mutual flux, ϕ_m, can be written as

$$v_1(t) = N_p \frac{d\phi_m}{dt} \tag{3.6}$$

and

$$v_2(t) = N_s \frac{d\phi_m}{dt} \tag{3.7}$$

Taking the ratio of $v_1(t)$ to $v_2(t)$ using the above equations gives

$$\frac{v_1(t)}{v_2(t)} = \left(N_p \frac{d\phi_m}{dt} \right) \left(\frac{1}{N_s} \frac{dt}{d\phi_m} \right) = \frac{N_p}{N_s} \tag{3.8}$$

As a result, the ratio of the primary voltage to the secondary voltage both caused by the mutual flux is equal to the turns ratio of the transformer. It is desired that transformers have very low leakage flux such that

$$\phi_m \gg \phi_{Lp}; \quad \phi_m \gg \phi_{Ls} \tag{3.9}$$

Then, Equation 3.8 reduces to

$$\frac{v_p(t)}{v_s(t)} \approx \frac{N_p}{N_s} \tag{3.10}$$

This is the characteristic of ideal transformers. The same derivation techniques can also be used to obtain the relation between the current on the primary and secondary sides as

$$\frac{i_{p(t)}}{i_{s(t)}} \approx \frac{N_s}{N_p} \tag{3.11}$$

The equivalent circuit of the transformer is shown in Figure 3.2. Copper losses are represented by R_p and R_s whereas core losses are represented by R_c. X_M represents the magnetization current model. The leakage flux is represented by the primary and secondary reactances, X_p. The analysis of the transformer is usually performed by referring primary sides to the secondary sides or secondary sides to the primary sides. The equivalent circuit of the transformer referred to the primary side shown in Figure 3.3a is defined using the number of turns $a = N_p/N_s$.

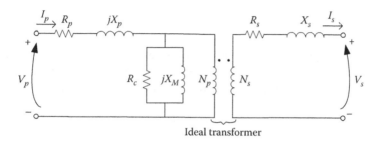

FIGURE 3.2 Equivalent circuit of the transformer.

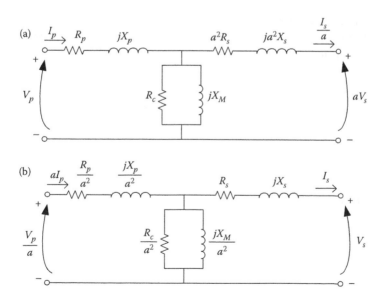

FIGURE 3.3 Equivalent circuit of the transformer referred to (a) the primary side and (b) the secondary side.

The mutual coupling between the primary side and the secondary side can be found by simplifying the equivalent circuit in Figure 3.2 by ignoring the losses as shown in Figure 3.4.

The mutual coupling is then defined as

$$M = k\sqrt{L_p L_s} \qquad (3.12)$$

where k is the coupling coefficient. L_p and L_s are defined by following the flux equations given by Equations 3.2 and 3.3 as

$$L_p = L_{1p} + L_{1l} \qquad (3.13)$$

$$L_s = L_{2s} + L_{2l} \qquad (3.14)$$

L_{1p} and L_{1l} are the primary and leakage inductances on the primary side of the transformer and L_{2s} and L_{2l} are the secondary and leakage inductances on the secondary side of the transformer. It is possible to reconfigure Figure 3.4 to include the leakage inductance as separate components as shown in Figure 3.5.

Then, the coupling coefficient can be expressed as

$$k = \sqrt{\frac{L_{1p} L_{2s}}{L_p L_s}} \qquad (3.15)$$

The analysis of the transformer can be facilitated by referring the transformer to either the primary side or the secondary side. The equivalent transformer circuit is referred to the primary side as shown in Figure 3.6.

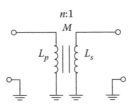

FIGURE 3.4 Equivalent circuit of the simplified transformer.

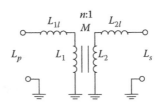

FIGURE 3.5 Simplified equivalent circuit with leakage inductances.

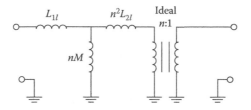

FIGURE 3.6 Equivalent transformer circuit referred to the primary side.

The inductances can be related to the mutual inductance and the number of turns as

$$L_p = L_{1l} + nM \tag{3.16}$$

$$M = \frac{L_p - L_{1l}}{n} \tag{3.17}$$

$$L_s = L_{2l} + \frac{M}{n} \tag{3.18}$$

Then

$$M = n(L_s - L_{2l}) \tag{3.19}$$

The values of the primary, secondary, and associated leakage inductances are measured by applying open and short circuit tests. The primary and secondary inductances, L_p and L_s, in Figure 3.5 are measured by leaving the terminals of the other side open. Leakage inductances are measured by short circuit test. The first short circuit test is done by short circuiting the secondary side of the transformer and measuring the inductance from the primary side of the transformer for the equivalent circuit in Figure 3.6 as shown in Figure 3.7.

The measurement shown in Figure 3.7 can be represented as

$$L_p' = L_{1l} + \left(\frac{n^2 M L_{2l}}{M + nL_{2l}} \right) \tag{3.20}$$

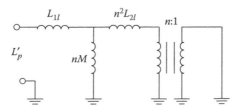

FIGURE 3.7 Secondary side leakage inductance measurement circuit.

Rearranging Equation 3.20 gives

$$L_p' = L_{1l} + \left(\frac{n^3(M/n)L_{2l}}{n(M/n) + nL_{2l}} \right) \tag{3.21}$$

Substitution of Equation 3.18 into Equation 3.21 gives

$$L_p' = L_{1l} + \left(\frac{n^3(L_s - L_{2l})L_{2l}}{n(L_s - L_{2l}) + nL_{2l}} \right) \tag{3.22}$$

or

$$L_p' = L_{1l} + \left(\frac{n^2(L_s - L_{2l})L_{2l}}{L_s} \right) \tag{3.23}$$

which can be written as

$$L_p' = L_{1l} + n^2 L_{2l} - \left(\frac{n^2 L_{2l}^2}{L_s} \right) \tag{3.24}$$

When we substitute Equation 3.19 into Equation 3.16, we obtain

$$L_p = L_{1l} + n^2(L_s - L_{2l}) \tag{3.25}$$

Then

$$L_{1l} = L_p - n^2(L_s - L_{2l}) \tag{3.26}$$

Substituting Equation 3.26 into Equation 3.24 gives

$$L_p' = L_p - n^2(L_s - L_{2l}) + n^2 L_{2l} - \left(\frac{n^2 L_{2l}^2}{L_s} \right) \tag{3.27}$$

or

$$\left(\frac{n^2 L_{2l}^2}{L_s} \right) - 2n^2 L_{2l} + (n^2 L_s - (L_p - L_p')) = 0 \tag{3.28}$$

Equation 3.28 is a quadratic equation in L_{2l}. All other parameters in the equation, including L_p, L_p', L_s, and n, are known parameters. Hence, leakage inductance on the secondary side of the transformer is found using Equation 3.28. L_p and L_s are measured with open circuit test measurements. The same procedure is repeated to find the leakage inductance on the primary side of the transformer by short circuiting the primary side as shown in Figure 3.8.

$$L_s' = L_{2l} + \left(\frac{ML_{1l}}{n^2 M + nL_{1l}} \right) \tag{3.29}$$

Substitution of Equation 3.17 into Equation 3.29 gives

$$L_s' = L_{2l} + \left(\frac{(L_p - L_{1l})L_{1l}}{n^2 L_p} \right) \tag{3.30}$$

or

$$L_s' = L_{2l} + \frac{L_{1l}}{n^2} - \left(\frac{L_{1l}^2}{n^2 L_p} \right) \tag{3.31}$$

Substituting Equation 3.17 into Equation 3.18 gives

$$L_s = L_{2l} + \frac{L_p - L_{1l}}{n^2} \tag{3.32}$$

From Equation 3.32, we obtain

$$L_{2l} = L_s - \frac{L_p - L_{1l}}{n^2} \tag{3.33}$$

Substitution of Equation 3.33 into Equation 3.31 gives

$$L_s' = L_s - \frac{L_p - L_{1l}}{n^2} + \frac{L_{1l}}{n^2} - \left(\frac{L_{1l}^2}{n^2 L_p} \right) \tag{3.34}$$

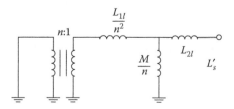

FIGURE 3.8 Primary side leakage inductance measurement circuit.

or

$$\left(\frac{L_{1l}^2}{n^2 L_p}\right) - \frac{2L_{1l}}{n^2} + \left(L_s' - L_s + \frac{L_p}{n^2}\right) = 0 \qquad (3.35)$$

Equation 3.35 is a quadratic equation in L_{1l}. All other parameters in the equation, including L_p, L_s', L_s, and n, are known parameters. Hence, leakage inductance on the primary side of the transformer is found using Equation 3.35. Once the leakage, primary, and secondary inductances are determined, the self-inductances are determined from Equations 3.13 and 3.14 as

$$L_{1p} = L_p - L_{1l} \qquad (3.36)$$

$$L_{2s} = L_s - L_{2l} \qquad (3.37)$$

The coupling coefficient is found from Equation 3.15. Mutual coupling can be then obtained from Equation 3.17 or Equation 3.19.

Design Example: Isolation Gate Transformer

It is required to design and characterize an isolation gate transformer to operate at 13.56 MHz. The current at the primary side is given to be 1.73 A_{rms}. The input power is 150 [W], and the voltage at the secondary side is desired to be 30 [V_{rms}]. The input impedance that will be presented is 50 Ω.

SOLUTION

On the basis of the given information, the voltage at the secondary side is 30 V. The voltage at the primary side is then equal to

$$V_p = \sqrt{150(50)} = 86.6 \ [V_{rms}]$$

The turns ratio is then calculated as

$$\frac{N_1}{N_2} = \frac{V_p}{V_s} = \frac{86.6}{30} = 2.89 \sim 3$$

Hence, the transformer with ratio 3:1 should satisfy the design specifications. Consider the single bead geometry given in Figure 3.9.

The cross-sectional area and magnetic path length are given as

$$A_e = 0.0996 \ cm^2, \quad l_e = 1.5 \ cm$$

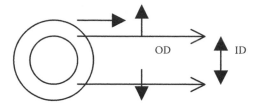

FIGURE 3.9 Single-bead geometry.

The permeability of the material is 125. In our design, we did not initially know the number of turns. So, as a rule of thumb, we targeted the impedance on the primary side as 10 times higher than the impedance that is interfaced.

$$Z = 2\pi f L = 2\pi(13.56 * 10^6) * L = 500\,\Omega$$

From this relation, the target inductance is 5.86 μH. When six beads are used, the total cross-sectional area becomes $A_{Te} = 6 \times A_e$ and the inductance value is 5.63 μH. Inductance with six beads is found from

$$L = \frac{0.4\pi\mu_i A_{Te}\,(\mathrm{cm}^2)N^2}{100 l_e\,(\mathrm{cm})}\,[\mu\mathrm{H}]$$

When this is placed back into the impedance formulation, we obtain

$$Z = 2\pi f L = 2\pi(13.56 * 10^6) * L = 479.8\,\Omega$$

A 3:1 transformer with six beads is realized using the bead configuration shown in Figure 3.10.

The size of the wires can be determined as follows. Since the current at the primary winding is 1.73 A, the minimum wire size should be 26 AWG. 26 AWG wire is able to handle 2.2 A if it is a solid wire. Since the current at the secondary winding is 5.19 A, the minimum wire size should be 22 AWG. 22 AWG is able to handle 5.5 A when solid wire is used. One of the design practice is to choose the wire insulation properly to prevent arcing between primary and secondary windings. This can be visualized using Figure 3.11.

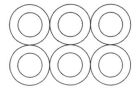

FIGURE 3.10 Six-bead transformer configuration.

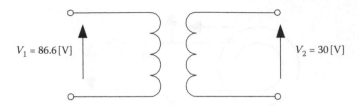

FIGURE 3.11 3:1 transformer illustration.

The voltage difference between the windings is $\Delta V = 86.6 - 30 = 56.6$ V. Hence, the breakdown voltage of the wires should be greater than 56.6 V. Teflon insulation used on the primary winding and enameled wire on the secondary winding has much more higher voltage rating than 56.6 V. The transformer is implemented in Figure 3.12. 14 AWG wire is used on the secondary side, which has much more higher current rating than the desired value. This transformer is used to drive metal oxide silicon field effect transistors (MOSFETs) and are connected between the gate and the source of the transistor.

Characterization of the transformer is done using open and short circuit tests as explained earlier. The open and short circuit test results for this transformer are shown below.

The measured values from Table 3.1 are

$$L_p = 5.15[\mu H], \quad L_s = 579.6[nH], \quad L'_s = 22[nH], \quad n = 3$$

From Equation 3.35

$$\left(\frac{L_{1l}^2}{n^2 L_p} \right) - \frac{2L_{1l}}{n^2} + \left(L'_s - L_s + \frac{L_p}{n^2} \right) = 0$$

FIGURE 3.12 Isolation gate transformer.

TABLE 3.1

Open and Short Circuit Test Results for Isolation Gate Transformer

Measured	Primary	Secondary	L
Primary	Open	Open	5.15 μH
Secondary	Open	Open	579.6 nH
Secondary	Short	Open	22 nH

or

$$\left(\frac{L_{1l}^2}{9 \left(5.15 \times 10^{-6} \right)} \right) - \frac{2 L_{1l}}{9} + \left(22 \times 10^{-9} - 579.6 \times 10^{-9} + \frac{5.15 \times 10^{-6}}{9} \right) = 0$$

The solution of the above equation gives the leakage inductance on the primary side as $L_{1l} = 66.22$ [nH]. The leakage inductance on the secondary side can be found using Equation 3.32 as

$$L_s = L_{2l} + \frac{L_p - L_{1l}}{n^2}$$

The leakage inductance on the secondary side is then equal to

$$L_{2l} = L_s - \frac{L_p - L_{1l}}{n^2} = 579.6 \, [\text{nH}] - \frac{5150 \, [\text{nH}] - 66.22 \, [\text{nH}]}{9} = 14.74 \, [\text{nH}]$$

Hence, the measured leakage inductances for the isolation gate transformer are

$$L_{1l} = 66.22 \, [\text{nH}] \quad \text{and} \quad L_{2l} = 14.74 \, [\text{nH}]$$

The self-inductances on the primary and secondary sides are found from

$$L_{1p} = L_p - L_{1l} = 5150 \, [\text{nH}] - 66.22 \, [\text{nH}] = 5083.78 \, [\text{nH}]$$

$$L_{2s} = L_s - L_{2l} = 579.6 \, [\text{nH}] - 14.74 \, [\text{nH}] = 564.86 \, [\text{nH}]$$

The coupling coefficient is obtained from

$$k = \sqrt{\frac{L_{1p} L_{2s}}{L_p L_s}} = \sqrt{\frac{(5083.78)(564.86)}{(5150)(579.6)}} = 0.98$$

The mutual inductance is then equal to

$$M = k\sqrt{L_p L_s} = 0.98\sqrt{(5150)(579.6)} = 1693.14[nH]$$

3.2 AUTOTRANSFORMER DESIGN

For some applications, it is desirable to change the voltage by a small amount. In such situations, it would be expensive to wind a transformer with two windings of approximately equal number of turns. An autotransformer, which is a transformer with only one winding, is used instead. The configurations of the step-down and step-up autotransformers are illustrated in Figure 3.13.

In Figure 3.13, V_C is the voltage across the common winding, I_C is the current through this coil, V_{SE} is the voltage across the series, and I_{SE} is the current through that coil. The voltages on high and low sides are called as V_H, I_H and V_L, I_L, respectively. In Figure 3.13a, the relations between voltages and currents and transformer ratios are given as

$$\frac{V_C}{V_{SE}} = \frac{N_C}{N_{SE}}$$ (3.38)

So

$$\frac{I_C}{I_{SE}} = \frac{N_{SE}}{N_C}$$ (3.39)

In addition, from the basic circuit theory

$$V_L = V_C$$
$$V_H = V_C + V_{SE}$$ (3.40)

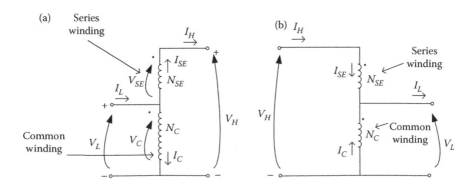

FIGURE 3.13   (a) Step-up transformer. (b) Step-down transformer.

and

$$I_L = I_C + I_{SE}$$
$$I_H = I_{SE} \tag{3.41}$$

Substituting Equation 3.38 into Equation 3.40 gives

$$V_H = V_C + \frac{N_{SE}}{N_C} V_C = V_L + \frac{N_{SE}}{N_C} V_L \tag{3.42}$$

or

$$\frac{V_L}{V_H} = \frac{N_C}{N_C + N_{SE}} \tag{3.43}$$

Equation 3.43 establishes the relation between voltages and turns ratio at high and low sides of the autotransformer. Applying the same procedure gives the relation between the current and the turns ratio at high and low sides of the autotransformer as

$$I_L = I_{SE} + \frac{N_{SE}}{N_C} I_{SE} = I_H + \frac{N_{SE}}{N_C} I_H \tag{3.44}$$

or

$$\frac{I_L}{I_H} = \frac{N_C + N_{SE}}{N_C} \tag{3.45}$$

Then, the impedance ratio is found by dividing Equation 3.43 to Equation 3.45 as

$$\frac{Z_L}{Z_H} = \left(\frac{N_C}{N_C + N_{SE}} \right)^2 \tag{3.46}$$

Design Example: Step-Down Autotransformer

It is desired to design autotransformers for RF amplifier load pull application at 2 MHz when the load presents (i) VSWR = 1.5:1, (ii) VSWR = 2:1, (iii) VSWR = 3:1, and (iv) VSWR = 5:1. It is given that the RF amplifier output power is 4000 W when terminated to 50 Ω load. Use K-type binocular ferrite core with $\mu = 125$, $l_e = 6.38$ cm, and $A_e = 2.22$ cm^2 in the construction of the transformers.

SOLUTION

The autotransformer configuration that will be constructed is illustrated in Figure 3.14 as explained earlier. In all the configurations, Z_H is the termination that will be

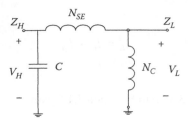

FIGURE 3.14 Step-down transformer for load pull applications.

interfaced with the amplifier output. The low-impedance port, $Z_L = 50\ \Omega$, is terminated to high-power load, which is also $50\ \Omega$. As a result, the low-impedance side of the transformer is matched with the amplifier. For load pulls, autotransformers can be used to make impedance transformation needed to provide the desired VSWR ratios. For instance, when VSWR = 1.5:1, it is required that $Z_H = 75\ \Omega$, when $Z_L = 50\ \Omega$. Our task is to find N_C and N_{SE} to give that impedance ratio. This relation is given by Equation 3.46.

The power and impedance information given in the design problem can be used to determine the winding that will be used in the transformer construction. The winding should be able to handle $V_{rms} = 447.2\ [V]$ and $I_{rms} = 8.94\ [A]$. Copper strip with 350 mil width and 10 mil thickness is a conservative choice that will be able to carry the amount of current calculated for VSWRs that are specified. The insulation for the copper strip is provided with 1 mil Teflon jacket to withstand the breakdown voltage.

   i. VSWR = 1.5:1

     VSWR = 1.5:1 can be obtained when $N_{SE} = 1$ and $N_C = 4$. The low-impedance side of the transformer is fixed and can be used as a reference. We target to have 5–10 times of the terminating impedance for inductance calculation. So, the impedance at the fixed impedance side is $Z = 10(50) = 500\ [\Omega]$. From the target impedance, we obtain the inductance value as

$$L \geq \frac{500}{2\pi(2 \times 10^6)} = 39.79\ [\mu H]$$

The number of turns required to obtain the inductance desired is found from

$$N^2 = \frac{39.79 \times 100 \times (6.38)}{0.4\pi(125)(2.22)} = 72.8 \rightarrow N > 8$$

Although, the number of turns, $N_C = 4$, is less than the minimum desired number of turns for the required inductance value, it is acceptable to use it due to lower operational frequency and very conservative choice of target impedance. When $N_C = 4$, the inductance value is obtained to be 8.75 [μH]. The autotransformer equivalent circuit that gives VSWR = 1.5:1 is shown in Figure 3.15. The compensation capacitor, C, which is connected in shunt at the high-impedance side, can be used when needed to reduce the leakage inductance.

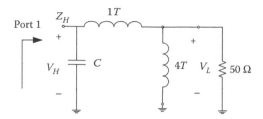

FIGURE 3.15   VSWR = 1.5:1 autotransformer.

On the basis of the turns ratio, the impedance at the high side is found from

$$Z_H = \left(\frac{N_C + N_{SE}}{N_C}\right)^2 Z_L = \left(\frac{5}{4}\right)^2 50 = 78.125 \; [\Omega]$$

The VSWR is then obtained as

$$\text{VSWR} = \frac{78.125}{50} = 1.56$$

This is very close to the required VSWR ratio and as a result we can construct the autotransformer based on the turns ratio that we determined. The constructed autotransformer is shown in Figure 3.16. Each autotransformer is constructed similarly with different number of turns.

The performance of the autotransformer is measured by the network analyzer from 1.8 to 8 MHz as shown in Figure 3.17. The measurement results show that the transformer performance at 2 MHz is in agreement with the calculated results as desired. The impedance measured at port 1 at 2 MHz is 77.815 Ω. The leakage reactance is significantly reduced with the shunt capacitor, 250 [pF]. The windings for N_C and N_{SE} are done separately with the copper strips and connections are made as illustrated in Figure 3.16.

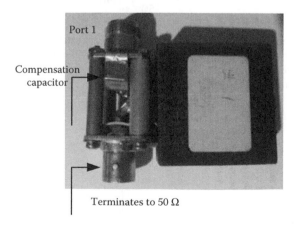

FIGURE 3.16   Constructed 1.5:1 autotransformer.

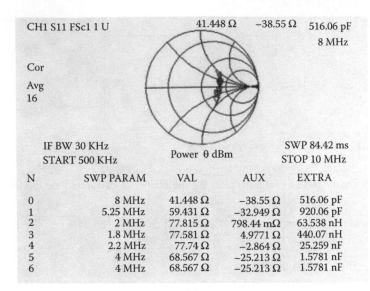

N	SWP PARAM	VAL	AUX	EXTRA
0	8 MHz	41.448 Ω	−38.55 Ω	516.06 pF
1	5.25 MHz	59.431 Ω	−32.949 Ω	920.06 pF
2	2 MHz	77.815 Ω	798.44 mΩ	63.538 nH
3	1.8 MHz	77.581 Ω	4.9771 Ω	440.07 nH
4	2.2 MHz	77.74 Ω	−2.864 Ω	25.259 nF
5	4 MHz	68.567 Ω	−25.213 Ω	1.5781 nF
6	4 MHz	68.567 Ω	−25.213 Ω	1.5781 nF

FIGURE 3.17   Measurement results of 1.5:1 autotransformer.

ii. VSWR = 2:1

The autotransformer configuration that gives VSWR = 2:1 is obtained with $N_{SE} = 2$ and $N_C = 5$ as shown in Figure 3.18. Then, the impedance at the high-impedance side is

$$Z_H = \left(\frac{N_C + N_{SE}}{N_C} \right)^2 Z_L = \left(\frac{7}{5} \right)^2 50 = 98 \ [\Omega]$$

This gives the VSWR ratio as

$$\mathrm{VSWR} = \frac{98}{50} = 1.96$$

The autotransformer is constructed and measured. The measurement results are illustrated in Figure 3.19. The measured impedance at the operational frequency, 2 MHz, is 97.408 [Ω]. The compensation capacitor in this configuration is 150 [pF].

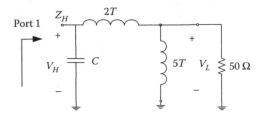

FIGURE 3.18   VSWR = 2:1 autotransformer.

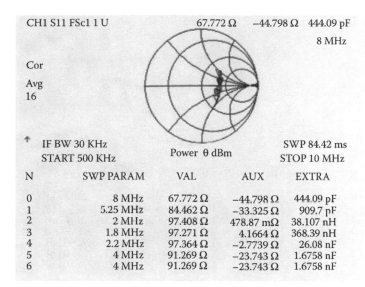

N	SWP PARAM	VAL	AUX	EXTRA
0	8 MHz	67.772 Ω	−44.798 Ω	444.09 pF
1	5.25 MHz	84.462 Ω	−33.325 Ω	909.7 pF
2	2 MHz	97.408 Ω	478.87 mΩ	38.107 nH
3	1.8 MHz	97.271 Ω	4.1664 Ω	368.39 nH
4	2.2 MHz	97.364 Ω	−2.7739 Ω	26.08 nF
5	4 MHz	91.269 Ω	−23.743 Ω	1.6758 nF
6	4 MHz	91.269 Ω	−23.743 Ω	1.6758 nF

FIGURE 3.19   Measurement results of 2:1 autotransformer.

iii. VSWR = 3:1

The autotransformer configuration that gives VSWR = 3:1 is obtained with $N_{SE} = 3$ and $N_C = 4$ as shown in Figure 3.20. Then, the impedance at the high-impedance side is

$$Z_H = \left(\frac{N_C + N_{SE}}{N_C}\right)^2 Z_L = \left(\frac{7}{4}\right)^2 50 = 153.125 \ [\Omega]$$

This gives the VSWR ratio as

$$VSWR = \frac{153.125}{50} = 3.06$$

The autotransformer is constructed and measured. The measurement results are illustrated in Figure 3.21. The measured impedance at the operational frequency, 2 MHz, is 152.39 [Ω]. The compensation capacitor in this configuration is 150 [pF].

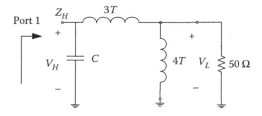

FIGURE 3.20   VSWR = 3:1 autotransformer.

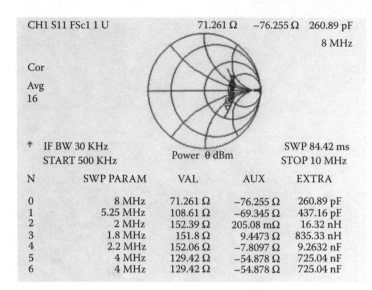

N	SWP PARAM	VAL	AUX	EXTRA
0	8 MHz	71.261 Ω	−76.255 Ω	260.89 pF
1	5.25 MHz	108.61 Ω	−69.345 Ω	437.16 pF
2	2 MHz	152.39 Ω	205.08 mΩ	16.32 nH
3	1.8 MHz	151.8 Ω	9.4473 Ω	835.33 nH
4	2.2 MHz	152.06 Ω	−7.8097 Ω	9.2632 nF
5	4 MHz	129.42 Ω	−54.878 Ω	725.04 nF
6	4 MHz	129.42 Ω	−54.878 Ω	725.04 nF

FIGURE 3.21   Measurement results of 3:1 autotransformer.

iv. VSWR = 5:1

The autotransformer configuration that gives VSWR = 5:1 is obtained with $N_{SE} = 5$ and $N_C = 4$ as shown in Figure 3.22. Then, the impedance at the high-impedance side is

$$Z_H = \left(\frac{N_C + N_{SE}}{N_C}\right)^2 Z_L = \left(\frac{9}{4}\right)^2 50 = 253.125 [\Omega]$$

This gives the VSWR ratio as

$$VSWR = \frac{253.125}{50} = 5.06$$

The autotransformer is constructed and measured. The measurement results are illustrated in Figure 3.23. The measured impedance at the operational frequency, 2 MHz, is 251.16 [Ω]. The compensation capacitor in this configuration is 86 [pF].

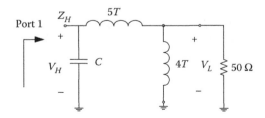

FIGURE 3.22   VSWR = 5:1 autotransformer.

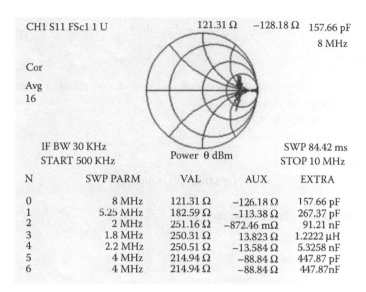

CH1 S11 FSc1 1 U   121.31 Ω   −128.18 Ω   157.66 pF

                                                                    8 MHz

Cor

Avg

16

IF BW 30 KHz   SWP 84.42 ms
START 500 KHz   Power θ dBm   STOP 10 MHz

N	SWP PARM	VAL	AUX	EXTRA
0	8 MHz	121.31 Ω	−126.18 Ω	157.66 pF
1	5.25 MHz	182.59 Ω	−113.38 Ω	267.37 pF
2	2 MHz	251.16 Ω	−872.46 mΩ	91.21 nF
3	1.8 MHz	250.31 Ω	13.823 Ω	1.2222 μH
4	2.2 MHz	250.51 Ω	−13.584 Ω	5.3258 nF
5	4 MHz	214.94 Ω	−88.84 Ω	447.87 pF
6	4 MHz	214.94 Ω	−88.84 Ω	447.87nF

FIGURE 3.23   Measurement results of 5:1 autotransformer.

3.3   TRANSMISSION LINE TRANSFORMERS

Transmission line transformers (TLTs) are constructed by winding pairs of parallel conducting wires separated by a uniform distance on the magnetic core. They have better frequency characteristics than conventional transformers for specifically higher-frequency applications [2–5]. It is a known fact that the interwinding capacitance of conventional transformers resonates with the leakage inductance and limits its high-frequency response. In TLTs, the coils are so arranged that the interwinding capacitance forms the characteristic impedance of the line and generates no resonances that cause major limitation in bandwidth. When the windings are spaced closely for TLTs, better coupling is achieved that results in broader bandwidth. The characteristics of two-wire transmission line (TL) wound on the core must be first determined before analyzing and dealing with TLTs. For a finite TL of length l terminated in an arbitrary load impedance Z_L shown in Figure 3.24, the input impedance looking into the TL at the input end of the line is given by the generalized transmission line theory

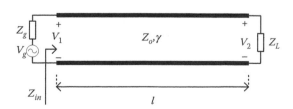

FIGURE 3.24   Circuit representation for transmission line analysis.

$$Z_i = Z_o \frac{Z_L + Z_o \tanh \gamma l}{Z_o + Z_L \tanh \gamma l} \tag{3.47}$$

where $\gamma = \alpha + j\beta$ is the propagation constant on the line. α is referred as attenuation and β is called the phase constant.

For a lossless transmission line

$$\gamma = j\beta \tag{3.48}$$

$$\tanh \gamma l = \tanh(j\beta l) = j\tan(\beta l) \tag{3.49}$$

Then, Equation 3.47 can be rewritten as

$$Z_i = Z_o \frac{Z_L + jZ_o \tan \beta l}{Z_o + jZ_L \tan \beta l} \tag{3.50}$$

The characteristic impedance of the TL is found using

$$Z_o = \sqrt{Z_{is} Z_{io}} \tag{3.51}$$

Z_{is} is the short circuit input impedance value when $Z_L \rightarrow 0$ in Equation 3.50. When $Z_L \rightarrow 0$ is substituted into Equation 3.50, we obtain

$$Z_{is} = jZ_o \tan \beta l \tag{3.52}$$

Z_{io} is the open circuit input impedance value when $Z_L \rightarrow \infty$ in Equation 3.50. When $Z_L \rightarrow \infty$ is substituted into Equation 3.50, we obtain

$$Z_{io} = -jZ_o \cot \beta l \tag{3.53}$$

As a result, the characteristic impedance of TL is independent of the length of the TL as shown by Equations 3.51 through 3.53. The schematic and electrical equivalent representation of TL is given in Figure 3.25.

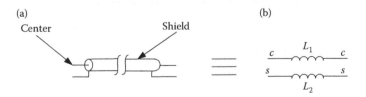

FIGURE 3.25   (a) Schematic representation. (b) Electrical equivalent.

TLT consists of two-wire lines that are usually implemented as twisted wire configuration. The characteristic impedance of the wire is determined based on the electrical properties of the insulation of the wire, its thickness, wire diameters, and so on. The practice to form the TLT is to use identical wires, which have the same characteristic impedances.

Design Example: Measuring Characteristic Impedance of Twisted Wire Transmission Line

Design a twisted wire TL configuration with 25 Ω characteristic impedance that will be able to handle 6 A current flow. Assume that the insulation thickness is 0.21 in. PTFE with $\varepsilon_r = 2.25$ and wall thickness $t = 0.045$ in is used as the insulation material to provide high-voltage breakdown. Calculate, measure, and confirm the characteristic impedance.

SOLUTION

The twisted wire configuration is illustrated in Figure 3.26.

18 AWG wire handles over 8 A when it is a single solid wire based on the information given in Table 2.2. When this wire is twisted with another 18 AWG, the current capacity doubles. Twisted wire TL characteristic impedance can be found using [6–12]

$$Z_0 = \frac{120}{\sqrt{\varepsilon_r}} \ln\left(\frac{2s}{d}\right) \, [\Omega] \tag{3.54}$$

where s is the separation distance, d is the wire diameter, and ε_r is the relative permittivity constant. The capacitance and inductance of the twisted wire are found from the following relations:

$$C = \frac{0.027815 \times 10^{-12} l}{\ln(2s/d)} \, [F] \tag{3.55}$$

$$L = 0.4 \times 10^{-9} l \ln\left(\frac{2s}{d}\right) \, [H] \tag{3.56}$$

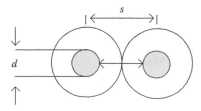

FIGURE 3.26 Twisted wire configuration.

All the lengths in Equations 3.55 and 3.56 are given in millimeters. On the basis of the given information

$$d = 1.024 \text{ mm}, \quad s = 2t = 2.286 \text{ mm}, \quad \text{and} \quad \varepsilon_r = 2.25$$

Substituting Equation 3.55 into Equation 3.54 gives the characteristic impedance as

$$Z_o = \frac{120}{\sqrt{2.25}} \ln\left(\frac{2(2.286)}{1.024}\right) = 119.6987 \, [\Omega]$$

The length of the wire is taken to be $l = 1$ ft = 30.48 cm. The capacitance and inductance of the wire are found from

$$L = 0.4 \times 10^{-9}(304.8)\ln\left(\frac{2(2.286)}{1.024}\right) = 182.42 \, [\text{nH}]$$

$$C = \frac{0.027815 \times 10^{-12}(304.8)}{\ln(2(2.286)/1.024)} = 12.75 \, [\text{pF}]$$

It can be proven that the characteristic impedance can also be found from

$$Z_o = \sqrt{\frac{L}{C}} = \sqrt{\frac{182.42 \times 10^{-9}}{12.75 \times 10^{-12}}} = 119.6186 \, [\Omega]$$

Now, we cut the 2×18 AWG wire in 1 ft length for twisting to form the TL as shown in Figure 3.27.

The measurement procedure begins with the visualization of the twisted wire as a TL as illustrated in Figure 3.27b. Assume that black (B) wire corresponds to the shield and red (R) wire corresponds to the center for the corresponding TL. This is similar to the configuration of a coaxial TL shown in Figure 3.27a. Using impedance analyzer, the capacitance, C, of the twisted wire is measured by leaving the terminals 3 and 4 open. The inductance, L, of the twisted wire is measured by shorting the terminals 3 and 4. When the experiment is performed at $f = 13.56$ MHz, the capacitance and inductance are measured to be

$$C = 13 \text{ pF} \quad \text{and} \quad L = 191 \text{ nH}$$

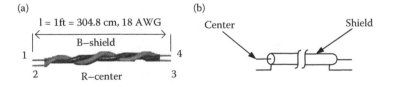

FIGURE 3.27 (a) 1 ft 18 AWG wire as transmission line to give 120 Ω characteristic impedance. (b) Equivalent schematic representation.

FIGURE 3.28 Parallel connection of twisted transmission lines to obtain 60 Ω characteristic impedance.

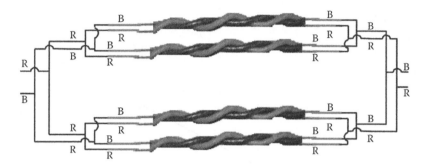

FIGURE 3.29 Parallel connection of twisted transmission lines to obtain 30 Ω characteristic impedance.

which leads to the characteristic impedance

$$Z_o = \sqrt{\frac{L}{C}} = \sqrt{\frac{191 * 10^{-9}}{13 * 10^{-12}}} = 121.21\,\Omega$$

The measured values are very close to the calculated values. It is also possible to obtain different values of characteristic impedances using the base model given in Figure 3.27a. For instance, the characteristic impedance that is half of the value measured and calculated previously can be obtained by connecting two of the twisted base models constructed using the 18 AWG wire pair in parallel as shown in Figure 3.28. The theoretical characteristic impedance for this model is found to be 59.8093 Ω ~60 Ω.

Now, if four of the twisted base models are constructed using the 18 AWG wire pair in parallel as shown in Figure 3.29, we obtain the theoretical characteristic impedance of 30 Ω, which is close to the desired impedance value that can be used in the application.

When the configuration in Figure 3.29 is measured with the impedance analyzer, the measured value is found to be $Z_o = 29.5$ Ω as expected. The chart giving the characteristic impedance of twisted wires for various AWG sizes versus wall thickness is given in Figure 3.30.

3.4 HIGH-CURRENT TRANSMISSION LINE TRANSFORMER

The high-current capacity for transformers can be obtained if windings are configured in the microstrip structure as illustrated in Figure 3.31.

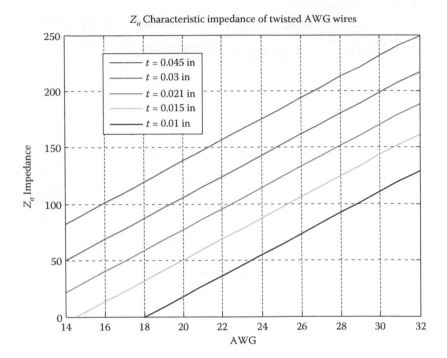

FIGURE 3.30 Characteristic impedance of twisted wires versus AWG sizes for various wall thicknesses.

As shown in Figure 3.31, the microstrip TL is formed with two separate copper strips as conductors, which represent ground and center conductors, respectively. A dielectric insulator such as Teflon is placed between two conductor strips. The constructed formation then becomes the microstrip TL configuration as illustrated in Figure 3.31. This structure is used for winding the core to implement TLT. In practice, the conductor strips and the dielectric insulator at the center are taken to be of equal width. The physical dimensions, including copper strip width and thickness, dielectric width, thickness, and permittivity constant, determine the characteristic impedance of TL. The characteristic impedance of the microstrip TL is found from

$$Z_o = \frac{60}{\sqrt{\varepsilon_{eff}}} \ln\left(\frac{8d}{W} + \frac{W}{4d}\right) \quad \text{for } \frac{W}{d} \le 1 \tag{3.57}$$

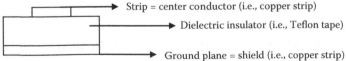

FIGURE 3.31 Microstrip configuration for transformers.

where

$$\varepsilon_{eff} = \frac{\varepsilon_r + 1}{2} + \frac{\varepsilon_r - 1}{2}\left[\left(1 + 12\frac{d}{W}\right)^{-(1/2)} + 0.04\left(1 - \frac{W}{d}\right)^2\right] \qquad (3.58)$$

or

$$Z_o = \frac{120\pi}{\sqrt{\varepsilon_{eff}}\left[W/d + 1.393 + (2/3)\ln\left(W/d + 1.444\right)\right]} \quad \text{for } \frac{W}{d} > 1 \qquad (3.59)$$

where

$$\varepsilon_{eff} = \frac{\varepsilon_r + 1}{2} + \frac{\varepsilon_r - 1}{2}\left(1 + 12\frac{d}{W}\right)^{-(1/2)} \qquad (3.60)$$

The accuracy of the formulation given by Equations 3.57 and 3.60 is increased by inclusion of the thickness of the strip. Any realizable conductor will have a finite thickness (t) causing fringing. The effect of nonzero conductor thickness can be approximated as an increase in the effective width (W_{eff}) of the conductor as follows:

$$W_{eff} = W + \frac{t}{\pi}\left(1 + \ln\frac{2x}{t}\right) \qquad (3.61)$$

where x can take on two different values:

$$x = d \quad \text{for } W > \frac{d}{(2\pi)} > 2t \qquad (3.62)$$

or

$$x = 2\pi W \quad \text{for } \frac{d}{(2\pi)} > W > 2t \qquad (3.63)$$

It is a design practice to provide insulation on the exterior of the windings to avoid any arcing due to high current flow. This can be provided by using a dielectric jacket such as Teflon. The new configuration with dielectric jacket implemented is illustrated in Figure 3.32.

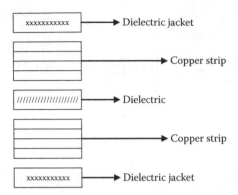

FIGURE 3.32   Configuration with dielectric jackets.

Design Example: Measuring Characteristic Impedance of High-Current Strip Transmission Line

Design a high-current microstrip TL-type winding for TLT with 25 Ω characteristic impedance, current capacity of 14 A, and 350 V minimum breakdown voltage. Obtain TL's frequency response between 0.5 and 6 MHz.

SOLUTION

Figure 3.33 shows the configuration that will give a minimum 14 A current capacity and well over 350 V breakdown voltage level.

The physical dimensions of the microstrip configuration in Figure 3.33 are $W = 300$ mil $= 0.00762$ m, $t = 40$ mil $= 0.001016$ m, $\varepsilon_r = 2.08$, and $d = 35$ mil $= 0.000889$ m.

The physical construction of the high-current TL winding is done as follows. 4×10 mil, 300 mil width copper strips are placed on the top and bottom of the configuration shown in Figure 3.33. 10 mil copper can carry a current of around 3.5 A. Hence, the total amount of current is 14 A for a 40-mil-thick copper strip, which gives the desired current rating. The breakdown voltage should be minimum

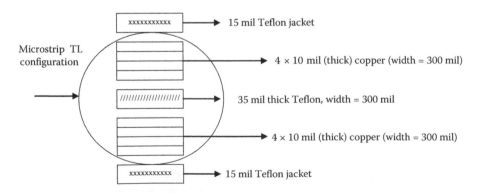

FIGURE 3.33   25 Ω characteristic impedance, minimum 14 A current-carrying capacity, and minimum 350 V breakdown voltage level.

350 V. There is a 25-mil-thick, 300-mil-wide Teflon insulation between the copper strip stack. 1 mil Teflon has a breakdown voltage of around 700 V. As a result, the thickness of Teflon guarantees the level of breakdown voltage needed. 15 mil Teflon jacket is placed around the configuration as shown in Figure 3.33 to prevent arcing. If higher voltage breakdown level is needed, it is a common practice to use 1 mil Teflon kapton tape for each copper strip to increase the breakdown voltage level. It is worth noting that the exterior Teflon jacket will slightly decrease the capacitance of the overall configuration and hence increase the characteristic impedance of the structure. The following MATLAB® script is used to calculate the characteristic impedance for high-current TL in Figure 3.33 without the effect of the Teflon jacket:

MATLAB Script: High-Current TL

```
% This program calculates the characteristic impedance of a
microstrip
% transmission line for given the thickness of the substrate,
width of
% the trace, relative permittivity of the substrate, and the
thickness
% of metallization.
%     d - Thickness of the substrate
%     W - Width of the trace
%     er - relative permittivity of the substrate
%     t - thickness of metallization

clear;
c = 3e8; %speed of light in free space

er = input('Relative Permittivity of Substrate (er) = ');
W = input('Width of the strip in m (W) = ');
d = input('Substrate Thickness in m (d) = ');
t = input('Conductor Thickness in m (t) = ');
l = input('Length of Line in m (l) = ');
% Calculate effective width
if (W > d/(2*pi))
    x = d;
else
    x = 2*pi*W;
end;
W_eff = W + t/pi*(1 + log(2*x/t));

% Calculate Zo
if (W/d < =1)
    % Effective permittivity
    epsilon_eff = (er + 1)/2 + ((er - 1)/2)*...
        ((1 + 12*(d/W_eff))^(-0.5) + 0.04*(1 - W_eff/d)^2);
    Zo = 60/sqrt(epsilon_eff)*log(8*d/W_eff + W_eff/(4*d));
    Vp = c./sqrt(epsilon_eff);
    C = (1./(Vp.*Zo)).*l;
    L = (Zo.^2).*C;
```

```
else
    epsilon_eff = (er + 1)/2 + ((er-1)/2)*...
        (1 + 12*(d/W_eff))^(-0.5);
    Zo = (120 * pi)/...
        (sqrt(epsilon_eff)*(W_eff/d + 1.393 + (2/3)*
log(W_eff/d + 1.444)));
    Vp = c./sqrt(epsilon_eff);
    C = (1./(Vp.*Zo)).*l;
    L = (Zo.^2).*C;
end;
display = ['The characterictic impedance is : ', num2str(Zo),'
Ohms']
display = ['The Capacitance is : ', num2str(C),' F']
display = ['The Inductance is : ', num2str(L),' H']
```

When the MATLAB script is executed, the following output is obtained. The physical dimensions are entered as input.

Relative permittivity of substrate (er) = 2.08
Width of the strip in m (W) = 0.00762
Substrate thickness in m (d) = 0.000889
Conductor thickness in m (t) = 0.001016
Length of line in m (l) = 0.3048
display =
The characterictic impedance is: 22.624 Ohm
The capacitance is: 6.1821e-011 F
The inductance is: 3.1643e-008 H

The TL using the dimensions given above is constructed and illustrated in Figure 3.34. The TL in Figure 3.34 is measured using the network analyzer to obtain its frequency response, including impedance, capacitance, and inductance values by leaving one of the terminals open and short, respectively. Smith chart plots giving the frequency response of high-current TL between 0.5 and 6 MHz are shown in Figures 3.35 and 3.36 with open and short tests, respectively.

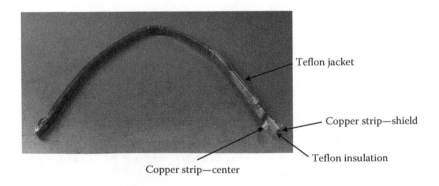

Teflon jacket

Copper strip—shield

Teflon insulation

Copper strip—center

FIGURE 3.34 Constructed high-current TL for winding.

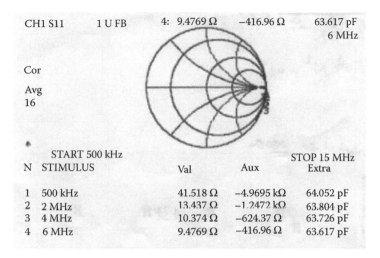

FIGURE 3.35 High-current TL frequency response with open test.

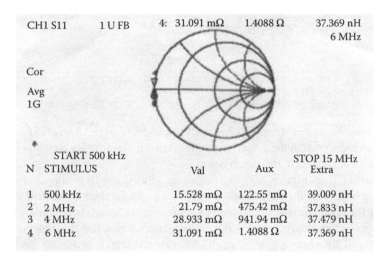

FIGURE 3.36 High-current TL frequency response with short test.

The capacitance and inductance values at 2 MHz are measured to be $C = 63.804$ [pF] and $L = 37.833$ [nH]. This leads to the measured characteristic impedance value as

$$Z_o = \sqrt{\frac{37.833 * 10^{-9}}{63.804 * 10^{-12}}} = 24.35\,\Omega$$

The measured characteristic impedance is slightly higher than the calculated value due to the external dielectric jacket as expected. The characteristic impedance

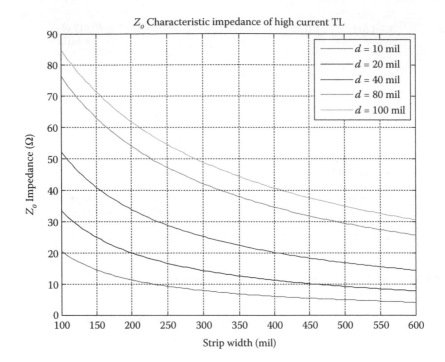

FIGURE 3.37 Characteristic impedance curves versus copper strip width for different dielectric thicknesses (Teflon, $\varepsilon_r = 2.08$).

curves versus the copper strip width when its thickness is 40 mil for various substrate thicknesses are shown in the graph below. Teflon with permittivity constant $\varepsilon_r = 2.08$ is used as the dielectric in the graph below.

As illustrated in Figure 3.37, as the strip width increases, the characteristic impedance lowers for the same substrate thickness. As the thickness of the substrate increases, the characteristic impedance increases for the same strip width. The characteristic impedance curves versus the copper strip width when the dielectric thickness is 40 mil for various conductor thicknesses are shown in Figure 3.38. Teflon with permittivity constant $\varepsilon_r = 2.08$ is used again as a substrate. As illustrated, the characteristic impedance of TL increases as the conductor thickness reduces for the same strip width.

The value of the characteristic impedance of the line that gives the best frequency response for TLT is found from

$$Z_o = \sqrt{Z_H Z_L} \tag{3.64}$$

where Z_H and Z_L are the high- and low-impedance values at the terminals of the transformer. As a result, the design practice for TLTs is to have the knowledge of the impedances at the terminals of the transformer and then determine the characteristic impedance of the TL using Equation 3.64. The next step is obtaining TL to give the

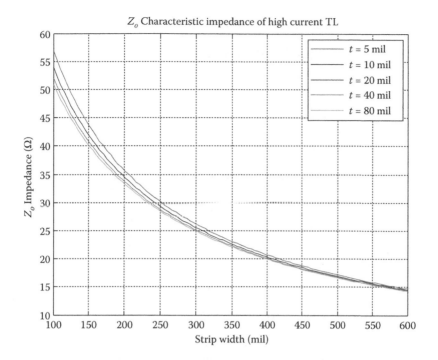

FIGURE 3.38 Characteristic impedance curves versus copper strip width for different conductor thicknesses (Teflon, $\varepsilon_r = 2.08$).

desired characteristic impedance obtained by Equation 3.64 using the two-wire line or microstrip TL configurations as described.

3.5 ARBITRARY TURNS RATIO TRANSMISSION LINE TRANSFORMERS

Arbitrary turns ratio can be obtained following the synthesis method described below. The idea is based on the use of Guanella's 1:1 TLT [1] as a basic block as shown in Figure 3.39.

The arbitrary ratio for the step-up transformer is formed by connecting the same ports of the basic block diagram to the transformer with $N_1:N_2$ transformer ratio when $N_1 < N_2$. Then, the resulting transformer ratio is $N_1:(N_1 + N_2)$ as illustrated in Figure 3.40. This type of connection is called parallel–series connection since we connect the basic block diagram and $N_1:N_2$ transformer in parallel at the input side

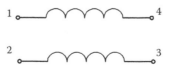

FIGURE 3.39 Basic block for arbitrary ratio TLT.

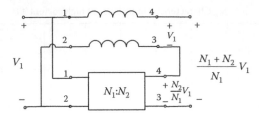

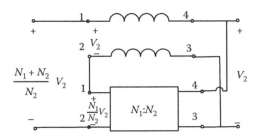

FIGURE 3.40 Arbitrary ratio for step-up TLT.

FIGURE 3.41 Arbitrary ratio for step-down TLT.

and in series at the output side. The arbitrary ratio for the step-down transformer is formed by connecting the basic block diagram in series at the input side and parallel at the output side as shown in Figure 3.41. Assuming $N_1 > N_2$, the resulting transformer ratio is $(N_1 + N_2):N_2$.

The synthesis procedure is applied as follows. Assume H to represent the high-impedance-side voltage ratio and L to represent the low-impedance-side voltage ratio of the transformer. Now, decompose TLT such that if H–L > L, then the TL is connected in series with the H–L side and in parallel with the L side. This is repeated until 1:1 transformer voltage ratio, which is Guanella's basic block, is achieved. The analysis of the arbitrary turns ratio TLT can be done using the equivalent circuit shown in Figure 3.42.

The primary voltage in the circuit can be found from

$$V_1 = V_s - I_1 R_s \tag{3.65}$$

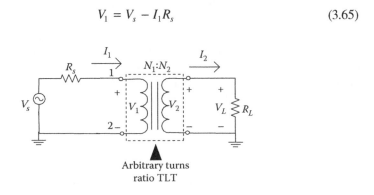

FIGURE 3.42 Equivalent circuit for arbitrary turns ratio TLT.

Using fundamental relations of the transformer

$$\frac{V_1}{V_2} = \frac{N_1}{N_2}, \quad \frac{I_1}{I_2} = \frac{N_2}{N_1}, \quad \text{and} \quad \frac{Z_1}{Z_2} = \left(\frac{N_1}{N_2}\right)^2 \tag{3.66}$$

The voltage on the secondary side of the transformer and load voltage is obtained as

$$V_2 = V_L = \frac{N_2}{N_1} V_1 = \frac{N_2}{N_1}(V_s - I_1 R_s) \tag{3.67}$$

From Kirchhoff's voltage law (KVL) on the secondary side of the transformer

$$R_L = \frac{V_2}{I_2} = \frac{(N_2/N_1)(V_s - I_1 R_s)}{I_1(N_1/N_2)} = \left(\frac{N_2}{N_1}\right)^2 \frac{(V_s - I_1 R_s)}{I_1} \tag{3.68}$$

The current on the primary side is then found by solving Equation 3.68 as

$$I_1 R_L = \left(\frac{N_2}{N_1}\right)^2 V_s - \left(\frac{N_2}{N_1}\right)^2 I_1 R_s \tag{3.69}$$

or

$$I_1\left(R_L + R_s\left(\frac{N_2}{N_1}\right)^2\right) = \left(\frac{N_2}{N_1}\right)^2 V_s \tag{3.70}$$

From Equation 3.70, I_1 is found as

$$I_1 = \frac{(N_2/N_1)^2 V_s}{(R_L + R_s(N_2/N_1)^2)} \tag{3.71}$$

Since the source voltage and source load impedances are known, the current on the primary side can be calculated from the relation given in Equation 3.71. Substitution of Equation 3.71 into Equation 3.65 gives the primary voltage as

$$V_1 = V_s - \frac{(N_2/N_1)^2 V_s}{R_L + R_s(N_2/N_1)^2} R_s = V_s\left(1 - \frac{(N_2/N_1)^2 R_s}{R_L + R_s(N_2/N_1)^2}\right) \tag{3.72}$$

The voltage and the current on the secondary side are found from Equation 3.66. In practice, the source voltage and impedance are required to be transformed and the terminated impedance are known parameters for the designer. Hence, the set of equations given by (3.65) and (3.72) can be used to determine the voltage, current, and power on either side of the transformer.

Example 1

a.   Obtain 3:5 voltage ratio using TLTs. Follow the synthesis technique and use Guanella's 1:1 transformer as a basic block. Use Pspice to confirm your results.
b.   Repeat part (a) and transform source impedance to the secondary side when the source voltage is 6 V. Compare the simulation results with the analytical results.

SOLUTION

Since the voltage turns ratio is given as 3:5, then $N_1 = 3$ and $N_2 = 5$. We implement the steps to construct the transformer as follows:

First step $\rightarrow V_{out} = V_{high}{:}V_{in} = V_{low} = N_2{:}N_1 = 5{:}3 \rightarrow (N_2 - N_1){:}N_1 = (5 - 3){:}3 = 2{:}3$
$= N_1'{:}N_2'$
Second step $\rightarrow V_{in} = V_{high}'{:}V_{out} = V_{low}' = N_2'{:}N_1' = 3{:}2 \rightarrow (N_2' - N_1'){:}N_1' = (3 - 2){:}$
$2 = 1{:}2 = N_1''{:}N_2''$
Third step $\rightarrow V_{out} = V_{high}''{:}V_{in} = V_{low}'' = N_2''{:}N_1'' = 2{:}1 \rightarrow (N_2'' - N_1''){:}N_1' = (2 - 1){:}$
$1 = 1{:}1 = N_1'''{:}N_2'''$

The implementation of the transformer begins from the last step. In the last step, $V_{in}{:}V_{out} = 1{:}2$ and it shows that the TLT is a step-up transformer. This corresponds to the transformer connection in Figure 3.40, parallel at the input and series at the output. TLs have 1:1 turns ratio to produce 1:2 turns ratio as illustrated in Figure 3.43 since N_1 and N_2 should be equal to 1 in Figure 3.40.

The second step shows that the TLT is now a step-down transformer. Therefore, the connection in Figure 3.41 is implemented with the aid of another TL as illustrated in Figure 3.44.

Now, we implement the first stage, which is another step-up transformer. The implementation of the first stage that shows the overall design of the TLT is given in Figure 3.45.

Pspice simulation is performed using the connection in Figure 3.45 as shown in Figure 3.46.

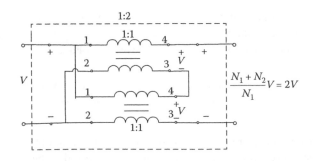

FIGURE 3.43   Implementation of third step: 1:2 ratio and step-up TLT.

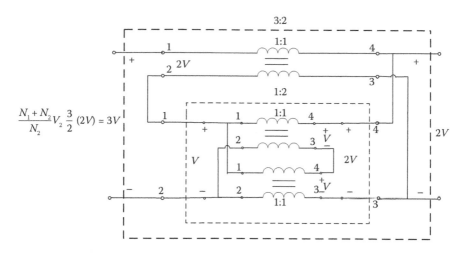

FIGURE 3.44 Implementation of second step: 3:2 ratio and step-down TLT.

The simulation results are given in Figure 3.47. The voltage level at each stage is illustrated and they match with the ones given in Figure 3.45. In the simulation of TLTs, it may sometimes require to use very small stability resistors to eliminate the inductor loops.

3:5 voltage ratio corresponds to 9:25 impedance ratio. The impedance at the secondary side is equal to

$$\frac{Z_1}{Z_2} = \left(\frac{N_1}{N_2}\right)^2 \quad \text{or} \quad Z_2 = \left(\frac{N_2}{N_1}\right)^2 Z_1 = \left(\frac{5}{3}\right)^2 9 = 25 \,[\Omega] \qquad (3.73)$$

Pspice schematics used for the simulation of the 9:25 impedance ratio TLT is shown in Figure 3.48.

The calculated current and voltage values are the peak values. The primary or the source current is found from Equation 3.71 as

$$I_1 = \frac{(N_2/N_1)^2 V_s}{R_L + R_s(N_2/N_1)^2} = \frac{16.67}{50} = 0.333\,[A] \qquad (3.74)$$

Then, the secondary side or the load current is found from

$$I_2 = \frac{N_1}{N_2} I_1 = (0.6)\,0.333 = 0.2\,[A] \qquad (3.75)$$

Pspice simulation results showing the primary and secondary currents are illustrated in Figure 3.49. They agree with the calculated values.

Substitution of Equation 3.75 into Equation 3.65 gives the primary voltage as

$$V_1 = V_s - I_1 R_s = 6 - (0.333)9 = 3\,[V] \qquad (3.76)$$

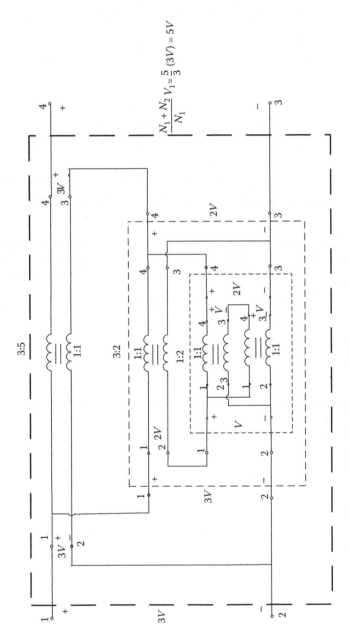

FIGURE 3.45 Implementation of first step: 3:5 ratio and step-up TLT.

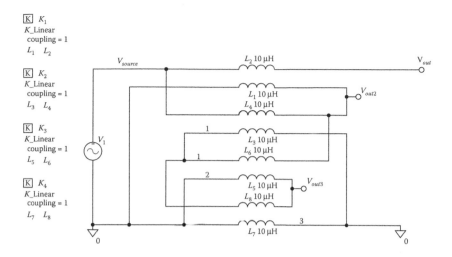

FIGURE 3.46   Pspice simulation of 3:5 ratio step-up TLT.

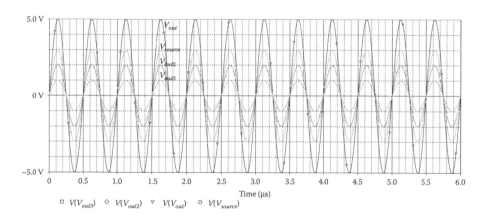

FIGURE 3.47   Pspice simulation results of 3:5 ratio step-up TLT.

Then, the secondary or the load voltage is found from Equation 3.66 as

$$V_2 = \left(\frac{N_2}{N_1}\right)V_1 = (1.667)3 = 5 \text{ [V]} \tag{3.77}$$

Pspice simulation results for the primary, secondary, and also intermediate voltages in the transformer are illustrated in Figure 3.50. They again agree with the calculated values.

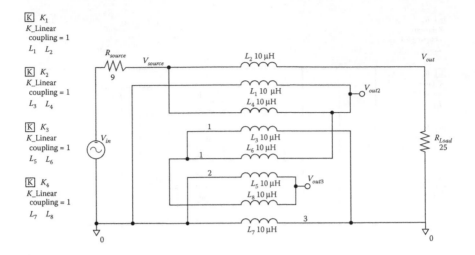

FIGURE 3.48   Pspice simulation of 9:25 impedance ratio TLT.

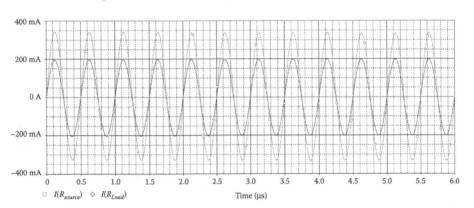

FIGURE 3.49   Pspice simulation results for currents.

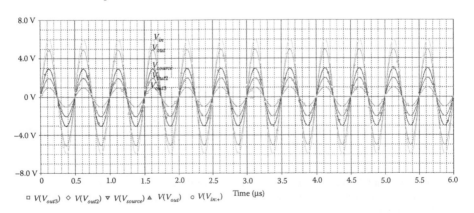

FIGURE 3.50   Pspice simulation results for voltages.

3.6 TLT DESIGN USING SERIES CONFIGURATION WITH FILARS: n^2:1 IMPEDANCE RATIO

TLTs can also be constructed using series connection with filar configuration. This is a practical implementation technique that is straightforward and expedites the design process. The most commonly used winding filar configurations are bifilar, trifilar, and quadrafilar configurations. Bifilar configuration has two parallel lines whereas trifilar and quadrafilar configurations have three and four parallel lines, respectively. The analysis of n-filar TLT configuration to give n^2:1 impedance ratio can be done using the circuit in Figure 3.51. All TLs used are identical in the construction.

The analysis of TLT in Figure 3.51 shows that

$$V_{in} = nV \tag{3.78}$$

$$V_{out} = -V \tag{3.79}$$

$$I_{in} = -\frac{I}{n} \tag{3.80}$$

$$I_{out} = I \tag{3.81}$$

Then

$$\frac{V_{out}}{V_{in}} = -\frac{1}{n} \tag{3.82}$$

$$\frac{I_{out}}{I_{in}} = -n \tag{3.83}$$

n represents the number of identical transmission lines used in the construction of the TLT. Current on the TL in the parallel branch, I_s, is found from

$$I_s = \frac{(n-1)Z}{nZ} I_{out} \tag{3.84}$$

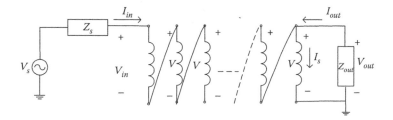

FIGURE 3.51 TLT design using series connection.

Dividing Equation 3.82 by Equation 3.83 gives

$$\left(\frac{V_{out}}{V_{in}}\right)\left(\frac{I_{in}}{I_{out}}\right) = \frac{1}{n^2} \tag{3.85}$$

which can be written as

$$\frac{Z_{out}}{Z_{in}} = \frac{1}{n^2}$$

or

$$Z_{out} = Z_{in}\left(\frac{1}{n^2}\right) \tag{3.86}$$

The equivalent circuit for the circuit shown in Figure 3.51 can be represented as shown in Figure 3.52.

The maximum power transfer for the circuit shown in Figure 3.52 is possible when

$$Z_{in} = Z_s \tag{3.87}$$

The current and the voltage are found as

$$I_{in} = \frac{V_s}{Z_s + Z_{in}} = \frac{V_s}{2Z_s} \tag{3.88}$$

and

$$V_{in} = \frac{V_s}{2} \tag{3.89}$$

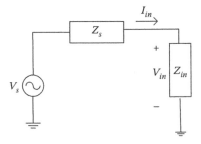

FIGURE 3.52 Simplified circuit for the circuit in Figure 3.51.

Example 2

Design a TLT to obtain 4:1 impedance ratio using series-type connection. Verify results using Pspice and obtain voltage and current waveforms. Source voltage, frequency, and impedance are given to be 10 V, 2 MHz, and 50 Ω, respectively.

SOLUTION

It is given that

$$V_s(t) = 10\sin(2\pi[2 \times 10^6]t) = 10\sin(4\pi10^6t), \quad Z_s = 50 \ [\Omega]$$

The configuration of the 4:1 impedance TLT or 2:1 voltage TLT is given in Figure 3.53. As a result, TLT is constructed in bifilar configuration.

A 4:1 impedance ratio or a 2:1 voltage ratio requires $n = 2$. That means two identical transmission lines have to be used for the construction of the TLT. For maximum power transfer

$$\frac{V_{in}}{I_{in}} = Z_{in} = Z_s = 50 \ [\Omega] \tag{3.90}$$

Using Equation 3.86

$$Z_{out} = Z_{in}\left(\frac{1}{n^2}\right) = 50\left(\frac{1}{4^2}\right) = 12.5 \ [\Omega] \tag{3.91}$$

The reactance of the TL is taken to be 5 times higher than the highest impedance termination. Then

$$Z = 5(50) = 250 \ [\Omega] = 2\pi fL \tag{3.92}$$

So, the inductance of each TL should be

$$L \geq \frac{250}{2\pi(2 \times 10^6)} = 19.89 \ [\mu H] \tag{3.93}$$

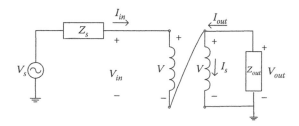

FIGURE 3.53 4:1 impedance ratio TLT design using series connection.

In practice, the core and the number of turns are determined from

$$L = \frac{0.4\pi\mu_i A_{Te}\ (\text{cm}^2)N^2}{100 l_e\ (\text{cm})}\ [\mu\text{H}] \tag{3.94}$$

Input current is found from Equation 3.88 as

$$I_{in} = \frac{V_s}{2Z_s} = \frac{10\sin(4\pi 10^6 t)}{2(50)} = 0.1\sin(4\pi 10^6 t)\ [\text{A}] \tag{3.95}$$

Output current is found from Equation 3.83 as

$$I_{out} = -n I_{in} = -2(0.1)\sin(4\pi 10^6 t) = 0.2\sin(4\pi 10^6 t + \pi)\ [\text{A}] \tag{3.96}$$

Current on the parallel branch is equal to

$$I_s = \frac{1}{2} I_{out} = 0.1\sin(4\pi 10^6 t + \pi)\ [\text{A}] \tag{3.97}$$

Similarly, the input and output voltages are found from Equations 3.89 and 3.82 as

$$V_{in} = \frac{V_s}{2} = \frac{10\sin(4\pi 10^6 t)}{2} = 5\sin(4\pi 10^6 t)\ [\text{V}] \tag{3.98}$$

and

$$V_{out} = -\frac{V_{in}}{n} = -\frac{5\sin(4\pi 10^6 t)}{2} = 2.5\sin(4\pi 10^6 t + \pi)\ [\text{V}] \tag{3.99}$$

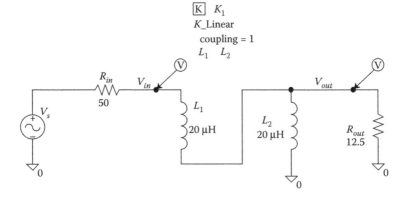

FIGURE 3.54 Pspice simulation of 4:1 impedance ratio TLT design using series connection.

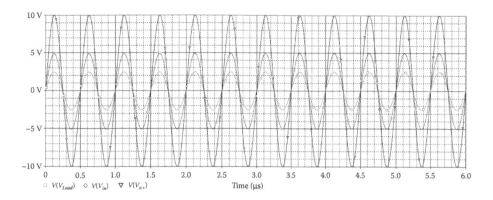

FIGURE 3.55 Pspice simulation results for voltage waveforms.

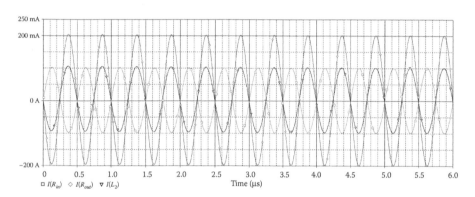

FIGURE 3.56 Pspice simulation results for current waveforms.

Owing to passive sign convention

$$V_{Load} = -V_{out} = 2.5 \sin(4\pi 10^6 t) \text{ [V]} \tag{3.100}$$

Pspice simulation circuit is given in Figure 3.54.

The voltage waveforms for input, source, and output voltages are given in Figure 3.55.

The current waveforms are given in Figure 3.56 for each branch.

The analytical and Pspice simulation results agree as illustrated.

3.6.1 TLT Design Using Series Configuration with Filars: $n^2{:}n_2^2$ Arbitrary Impedance Ratio

Arbitrary turns ratio TLT using filar configuration that is constructed with series connection of transmission lines are illustrated in Figure 3.57. Each TL in the configuration is assumed to be identical. If there are n identical transmission lines, and the structure is tapped from n_2th transmission line, then the overall TLT gives $n^2{:}n_2^2$ impedance ratio or $n{:}n_2$ voltage ratio.

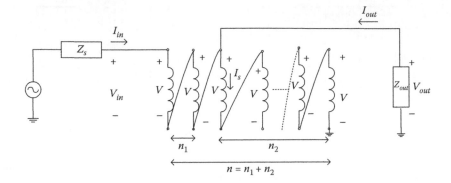

FIGURE 3.57   Arbitrary turns ratio TLT design using series connection.

The analysis of TLT in Figure 3.57 shows that

$$V_{in} = nV \tag{3.101}$$

$$V_{out} = -n_2 V \tag{3.102}$$

$$I_{in} = -\frac{n_2 I}{n} \tag{3.103}$$

$$I_{out} = I \tag{3.104}$$

Then

$$\frac{V_{out}}{V_{in}} = -\frac{n_2}{n} \tag{3.105}$$

$$\frac{I_{out}}{I_{in}} = -\frac{n}{n_2} \tag{3.106}$$

Current on the TL in the parallel branch, I_s, is found from

$$I_s = \frac{n_1}{n} I_{out} \tag{3.107}$$

Dividing Equation 3.105 by Equation 3.106 gives

$$\left(\frac{V_{out}}{V_{in}}\right)\left(\frac{I_{in}}{I_{out}}\right) = \frac{n_2^2}{n^2} \tag{3.108}$$

which can be written as

$$\frac{Z_{out}}{Z_{in}} = \frac{n_2^2}{n^2} \quad \text{or} \quad Z_{out} = Z_{in}\left(\frac{n_2^2}{n^2}\right) \tag{3.109}$$

Example 3

Design a TLT to obtain 9:4 impedance ratio using series-type connection. Verify the results using Pspice and obtain voltage waveforms and current waveforms. Source voltage, frequency, and impedance are given to be 10 V, 2 MHz, and 50 Ω, respectively.

SOLUTION

The source voltage is given as

$$V_s(t) = 10\sin(2\pi[2 \times 10^6]t) = 10\sin(4\pi10^6t), \quad Z_{in} = Z_s = 50\ [\Omega]$$

9:4 impedance ratio requires 3:2 voltage ratio. As a result, three identical transmission lines have to be used to construct the TL transformer. So, this is a trifilar TLT configuration with $n = 3$ and $n_2 = 2$. The output impedance is found from Equation 3.109 as

$$Z_{out} = Z_{in}\left(\frac{n_2^2}{n^2}\right) = 50\left(\frac{4}{9}\right) = 22.22\ [\Omega]$$

The inductance of each TL is calculated to be $L \geq 19.89$ [μH]. Input current is obtained as

$$I_{in} = \frac{V_s}{2Z_s} = \frac{10\sin(4\pi10^6t)}{2(50)} = 0.1\sin(4\pi10^6t)\ [A]$$

Output current is calculated from Equation 3.106 as

$$I_{out} = -\frac{n}{n_2}I_{in} = -\frac{3}{2}(0.1)\sin(4\pi10^6t) = 0.15\sin(4\pi10^6t + \pi)\ [A]$$

Current on the parallel branch is equal to

$$I_s = \frac{n_1}{n}I_{out} = \frac{1}{3}0.15\sin(4\pi10^6t + \pi) = 0.05\sin(4\pi10^6t + \pi)\ [A]$$

Input and output voltage are found from Equations 3.105 and 3.82 as

$$V_{in} = \frac{V_s}{2} = \frac{10\sin(4\pi10^6t)}{2} = 5\sin(4\pi10^6t)\ [V]$$

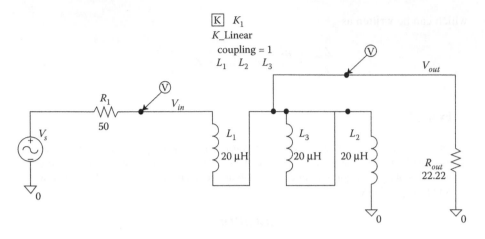

FIGURE 3.58 Pspice simulation of 3:2 impedance ratio TLT design using series connection.

and

$$V_{out} = -\frac{n_2}{n}V_{in} = -\frac{2}{3}5\sin(4\pi10^6 t) = 3.33\sin(4\pi10^6 t + \pi)[\text{V}]$$

So

$$V_{Load} = -V_{out} = 3.33\sin(4\pi10^6 t)[\text{V}]$$

The trifilar 3:2 turns ratio TLT is simulated using Pspice as shown in Figure 3.58. The voltage waveforms are shown in Figure 3.59.

As seen from Figures 3.59 and 3.60, the results agree with the analytical results. After identification of the number of turns, inductance value, and termination impedances, the only item in the design of TLT using the series connection with filar configuration is to determine the type of the core that will be used. The design examples will detail the determination of the core. The current waveforms are shown in Figure 3.60.

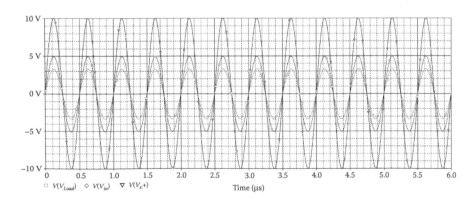

FIGURE 3.59 Pspice simulation results for voltage waveforms.

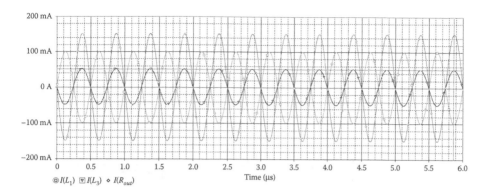

FIGURE 3.60 Pspice simulation results for current waveforms.

Design Example: Implementation of TLT Using Series Connection

Design a three-filar impedance TLT with 9:1 impedance ratio at 13.56 MHz. Eliminate the leakage inductance by using compensation techniques.

SOLUTION

A 9:1 impedance ratio requires 3:1 voltage turns ratio. Ferronics core with manufacturer part number 11-260 is selected to be used in the construction of the transformer. Core is a K-type material with initial permeability, $\mu_i = 125$. This is basically a ferrite bead in toroidal shape, which is used in multiple-bead configuration to increase the saturation flux. The details about the physical dimensions of several Ferronics toroid cores, including the one that is selected for this application, are given in Table 3.2.

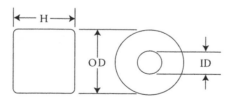

The configuration of the TLT is given in Figure 3.61.

Ten beads are stacked together, five on one side and five on the other. It is given in Table 3.2 that $l_{eTot} = l_{eT} = 2.95$ cm and $A_e = 0.129$ cm². So, $A_{eT} = 10 * A_e = 1.29$ cm². If we choose the higher impedance side to be 50 Ω, the required inductance value is found from

$$Z = 10(50) = 500 \ [\Omega] = 2\pi f L$$

So, the inductance is

$$L \geq \frac{225}{2\pi(13.56 \times 10^6)} = 5.87 \ [\mu H]$$

TABLE 3.2
Ferronics Toroid Core Physical Dimensions

| Part Number | | Physical Dimensions | | | | | | Effective Dimensions | |
Uncoated	Coated	OD Inches	OD Millimeters	ID Inches	ID Millimeters	H Inches	H Millimeters	A_e (cm²)	I_e (cm)
11-122	11-622	0.230 ±0.005	5.84 ±0.13	0.120 ±0.005	3.05 ±0.13	0.120 ±0.005	3.05 ±0.13	0.0411	1.30
11-160	11-660	0.300 ±0.006	7.62 ±0.15	0.125 ±0.005	3.18 ±0.13	0.188 ±0.005	4.78 ±0.13	0.0996	1.5
11-170	11-670	0.354 ±0.008	9.00 ±0.20	0.236 ±0.007	6.00 ±0.18	0.118 ±0.004	3.00 ±0.10	0.0443	2.29
11-220	11-720	0.375 ±0.007	9.53 ±0.18	0.187 ±0.005	4.75 ±0.13	0.125 ±0.005	3.18 ±0.13	0.0728	2.07
11-260	11-759	0.500 ±0.010	12.70 ±0.25	0.281 ±0.007	7.14 ±0.18	0.188 ±0.005	4.78 ±0.13	0.129	2.95
	11-760								
11-261	11-761	0.500 ±0.010	12.70 ±0.25	0.281 ±0.007	7.14 ±0.18	0.250 ±0.007	6.35 ±0.18	0.172	2.95
11-251	11-751	0.500 ±0.010	12.70 ±0.25	0.312 ±0.008	7.92 ±0.20	0.125 ±0.005	3.18 ±0.13	0.0744	3.12
11-250	11-750	0.500 ±0.012	12.70 ±0.25	0.312 ±0.008	7.92 ±0.20	0.250 ±0.007	6.35 ±0.18	0.149	3.12
11-247	11-747	0.551 ±0.012	14.00 ±0.30	0.354 ±0.008	9.00 ±0.20	0.197 ±0.005	5.00 ±0.13	0.123	3.50
11-270	11-770	0.625 ±0.014	15.88 ±0.36	0.350 ±0.008	8.89 ±0.20	0.185 ±0.005	4.70 ±0.13	0.160	3.68
11-280	11-780	0.870 ±0.017	22.10 ±0.43	0.540 ±0.008	13.72 ±0.20	0.250 ±0.007	6.35 ±0.18	0.261	5.42
11-282	11-782	0.870 ±0.017	22.10 ±0.43	0.540 ±0.012	13.72 ±0.30	0.500 ±0.013	12.70 ±0.30	0.522	5.42
11-295	11-795	0.906 ±0.020	23.00 ±0.51	0.551 ±0.012	14.00 ±0.30	0.276 ±0.007	7.00 ±0.18	0.310	5.58

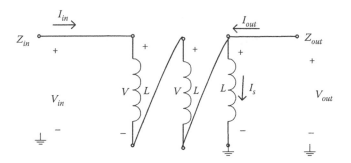

FIGURE 3.61 9:1 impedance ratio TLT design using series connection.

The number of turns required to obtain the inductance calculated is found from

$$N^2 = \frac{5.87 \times 100 \times (2.95)}{0.4\pi(125)(1.29)} = 8.54 \rightarrow N \sim 3$$

The constructed three-filar TLT is shown in Figure 3.62. A 9:1 impedance ratio TLT is tested using the network analyzer with open test as shown in Figure 3.63. The open circuit test results for this transformer are shown in Table 3.3.

The measured impedance ratio is 8.43:1, which is very close to the 9:1 ratio. The self-resonant frequency is desired to be away from the operational frequency. Let us check our TLT with load impedance, which is equal to 2.5 Ω. So, the 2.5 Ω resistor is connected to port 2 as shown in Figure 3.64 and impedance is measured from port 1.

The measured impedance at port 1 is

$$Z_1 = [21.552 + j22.225]\ \Omega$$

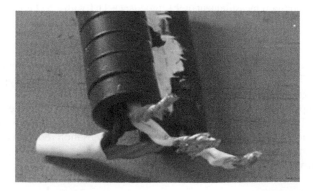

FIGURE 3.62 Constructed three-filar, 9:1 impedance ratio TLT with K-type material.

TABLE 3.3

Open Circuit Test Results 9:1 Impedance Ratio TLT

Measured	Port 1	Port 2	L
Primary	Open	Open	6.05 μH
Secondary	Open	Open	747 nH

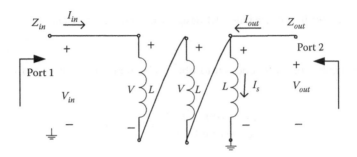

FIGURE 3.63   Open circuit measurement of 9:1 impedance TLT.

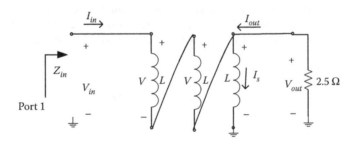

FIGURE 3.64   One-port measurement of 9:1 impedance TLT with 2.5 Ω impedance at port 2.

The leakage reactance $j22.225$ Ω needs to be eliminated using compensation techniques. This can be done by tuning TLT with lumped element capacitors as shown in Figure 3.65.

The compensation is done at the input and output of TLT by connecting shunt capacitors. The value and connection of the lumped element depends on the leakage inductance measured. In our case, the compensation reduced the leakage reactance impedance significantly when impedance is measured at port 1. The measured impedance at port 1 with the compensation is

$$Z_1 = [20.288 + j6.8584] \ \Omega$$

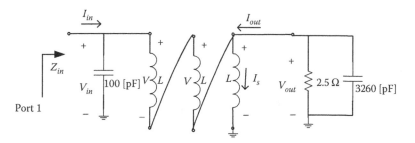

FIGURE 3.65 One-port measurement of 9:1 impedance TLT with 2.5 Ω impedance at port 2 using compensation.

3.7 ANALYSIS AND DESIGN OF BALUN USING TLT

TLT can be implemented as phase inverter or balun using the basic TLT block, Guanella's 1:1 transformer. The balun implementation is illustrated in Figure 3.66.

The analysis of the balun is performed by applying KVL around loops, KVL_1 and KVL_2, as illustrated in the figure. In Figure 3.66,

$$L_1 = L_2 = L \tag{3.110}$$

$$I_1 = -I_2 = I \tag{3.111}$$

$$V_1 = -V_2 = V \tag{3.112}$$

From KVL_1 and KVL_2

$$V_s - IR_s - V - \frac{V_L}{2} = 0 \tag{3.113}$$

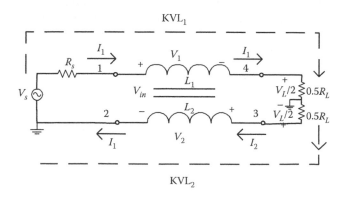

FIGURE 3.66 Implementation of balun.

$$\frac{V_L}{2} - V = 0 \tag{3.114}$$

Since

$$\frac{V_L}{2} = 0.5 R_L I \quad \text{then } I = \frac{V_L}{R_L} \tag{3.115}$$

Substituting Equations 3.114 and 3.115 into Equation 3.113 gives

$$V_s - \frac{V_L R_s}{R_L} - V_L = 0 \quad \text{or} \quad V_s = V_L \left(\frac{R_s + R_L}{R_L} \right) \tag{3.116}$$

As a result, when $R_s = R_L$, the output voltage is

$$V_s = V_L \left(\frac{2R_s}{R_s} \right) \quad \text{or} \quad V_L = \frac{V_s}{2} \tag{3.117}$$

which leads to

$$V_{R_L/2} = \frac{V_L}{2} = \frac{V_s}{4} = \frac{V_{in}}{2} \tag{3.118}$$

and the voltage across the windings of the balun is also

$$V = \frac{V_L}{2} = \frac{V_s}{4} \tag{3.119}$$

To have the magnetizing current to produce the amount of flux needed to couple the windings of the balun, the inductance of each winding should be set to at least 5 times higher than the terminating impedance. The circuit in Figure 3.66 is simulated with Pspice as shown in Figure 3.67. The results confirm the relations shown by Equations 3.110 through 3.119.

The netlist of the simulated circuit is

```
Kn_K1           L_L1 L_L2 1
R_Rload1        0 Vload1 25
R_Rload2        Vload2 0 25
V_Vs            Vsource 0 + SIN 0 10 2meg 0 0 0
R_Rs            Vsource Vin 50
L_L1            Vin Vload1 10uH
L_L2            0 Vload2 10uH
.Tran 0ns 10us 0 1n
.Probe
.End
```

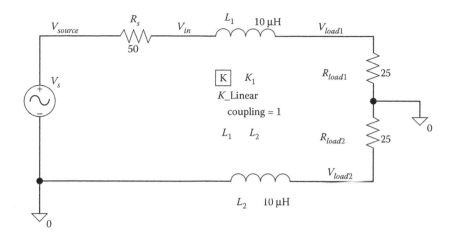

FIGURE 3.67 Pspice simulation of balun.

Using Probe in Pspice, input and output voltages are shown Figure 3.68.

Analytical and simulation results are in agreement as illustrated in Figure 3.68. The balun in Pspice can also be implemented using a TL component as shown in Figure 3.69.

The netlist of this circuit is

```
R_Rload1      0 Vload1 25
R_Rload2      Vload2 0 25
V_Vs          Vsource 0 + SIN 0 10 2meg 0 0 0
T_T1          Vin 0 Vload1 Vload2 Z0 = 50 F = 2meg NL = 0.03
R_Rs          Vsource Vin 50
.Tran 0ns 10us 0 1n
.Probe
.End
```

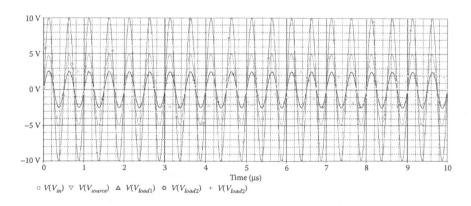

FIGURE 3.68 Pspice simulation results for balun.

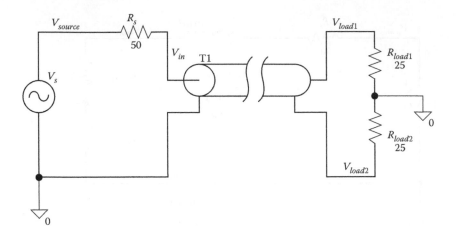

FIGURE 3.69   Alternative implementation of balun in Pspice.

The TL component has the following properties shown in the window below.

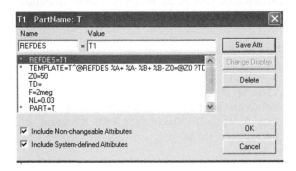

The critical component property NL or the normalized length in Pspice is calculated using

$$NL = \frac{f_{op}l}{v_p} \tag{3.120}$$

where l is the physical length of the line, v_p is the phase velocity, and f_{op} is the operational frequency. The simulation results are given in Figure 3.70 and are identical to the ones shown in Figure 3.68.

Design Example: Implementation of Input Balun

Design a balun to feed the inputs of a push–pull pair of RF amplifier, which has 25 Ω interface impedance at 13.56 MHz. The total power flowing into the balun

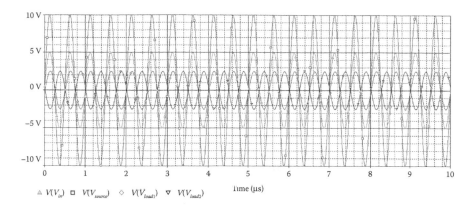

FIGURE 3.70 Pspice simulation results for alternative balun implementation.

is set to be 150 W to provide the required amount of power at the inputs of each side of the push–pull pair.

SOLUTION

On the basis of the given information, the equivalent circuit of the input balun can be illustrated as shown in Figure 3.71.

For maximum power transfer from Equation 3.116, $R_L = R_s$. It is given that 0.5 $R_L = 25$. Then

$$R_L = R_s = 50\ [\Omega]$$

It is also given that $P_{in} = 150$ W. This requires the operational rms voltage and current values at the input to be

$$P_{in} = 150\ \text{W},\quad Z_{in} = 50\ \Omega,\quad V_{in,rms} = 86.6\ \text{V},\quad \text{and}\quad I_{in,rms} = 1.73\ \text{A}$$

At the output of the balun, using Equation 3.118, we have

$$P_{out,1} = 75\ \text{W},\quad Z_{out,1} = 25\ \Omega,\quad V_{out,rms,1} = 43.3\ \text{V},\quad \text{and}\quad I_{out,rms,1} = 1.73\ \text{A}$$

The magnetic core that will be used in the design of the balun will be selected based on the impedance and flux density requirement. The rule of thumb in the

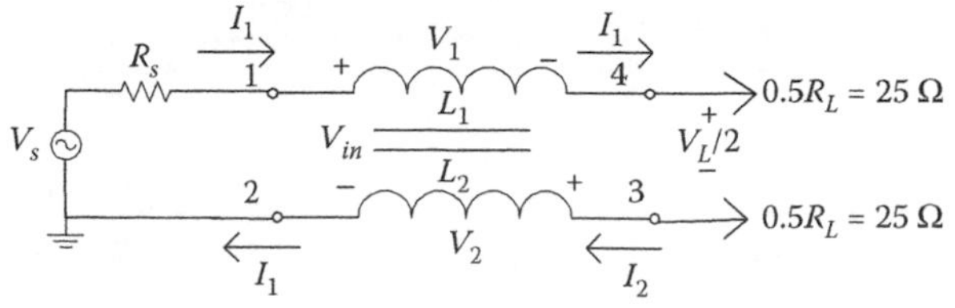

FIGURE 3.71 The equivalent circuit of input balun.

design of the balun is to have the impedance value at least 5–10 times higher than the highest impedance termination at any of its ports. The highest impedance exists at the input port of the balun and it is equal to 50 Ω. The target impedance value for the balun is then

$$Z_{req} = 2\pi f L = 2\pi(13.56 \times 10^6) \times L = 500\,\Omega$$

Then

$$L \geq \frac{500}{2\pi(13.56 \times 10^6)} \quad \text{or} \quad L \geq 5.868\,\mu H$$

The inductance is found from Equation 2.16a as

$$L = \frac{4\pi N^2 \mu_i A_{Tc}}{l_e} \quad [nH]$$

We can now find the number of turns, N, if we know the characteristics of the core we will be using. We determine to use the core material manufactured by Ferronics with part number 11-260. This is K-type material with a permeability of 125, $\mu_i = 124$, OD = 1.27 cm, ID = 0.714 cm, and $h = 0.478$ cm. The saturation flux density at 10 MHz from Table 2.9 is given as 3200 [G]. This value is expected to drop significantly as the frequency increases. We will be using 10 cores by stacking five of them on each side as shown in Figure 3.72.

Substitution of the known values into Equation 2.16a gives the number of turns as

$$N = \sqrt{\frac{L \times l_e}{4\pi \mu_i A_{Tc}}} = \sqrt{\frac{5868 \times 3.03}{4\pi \times 125 \times 10 \times 0.132}} = 2.92 \sim 3\,\text{turns}$$

The operational flux density B_{op} can be found using Equation 2.20 as

$$B_{op} = \frac{V_{rms} \times 10^8}{4.44 f N A_{Tc}} = \frac{86.6 \times 10^8}{4.44 \times 13.56 \times 10^6 \times 2.92 \times 10 \times 0.132} = 37.07\,[\text{Gauss}]$$

FIGURE 3.72   Construction of the balun using multiple cores.

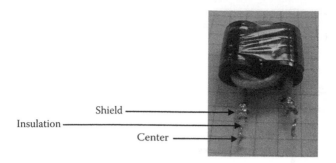

Insulation ─────────

Shield ─────────

Center ─────────

FIGURE 3.73 Balun that is constructed.

The operational flux density is well below the saturation flux density. It is also worth noting that increasing the number of cores reduces the operational flux density. RG-188 coaxial cable is selected for winding since it is capable of carrying the current required and has 50 Ω characteristic impedance.

The final configuration of the balun that is constructed is shown in Figure 3.73. MATLAB GUI program has been developed to design the balun for the desired operating conditions, including power and input impedance requirement. GUI output is illustrated in Figure 3.74 and its results confirm the previously calculated values. In the program, the physical dimensions of the selected core to obtain the desired inductance value are entered as input parameters. Initially, the desired inductance value is not known. However, the only known parameters for the balun design are input power, input impedance, and operational frequency. The program lets the user enter these input parameters and outputs L, V_{in}, V_{out}, I_{in}, I_{out}, and Z_{out}. Once L, which is the target inductance value, is known, the core parameters can now be entered and compared with the target inductance value. The calculated values are based on the rounded number of turns, N, which would be used in the construction of the balun. Operational flux density using the rounded number of turns is also calculated and illustrated in the program.

The final balun configuration illustrated in Figure 3.74 is constructed with three turns. When $N = 3$, using the inductance formulation, inductance is found to be 6.16 [μH]. The inductance of the balun is measured using HP 4191A, an RF impedance analyzer. The setup to measure the inductance of the balun is shown in

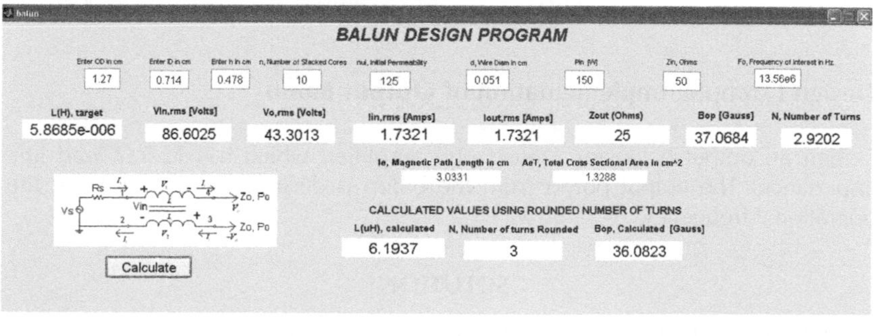

FIGURE 3.74 MATLAB GUI program to design balun.

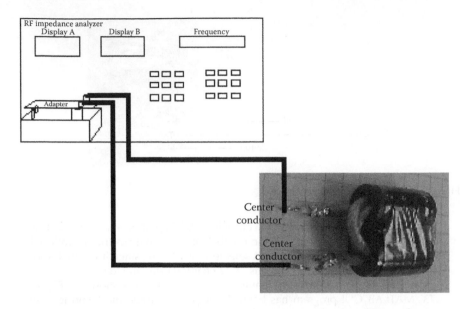

FIGURE 3.75   Measurement setup for inductance measurement of the input balun.

Figure 3.75. The self-inductance value is measured to be 6.4 [μH] and is close to the calculated value.

The frequency response of the input balun is measured with HP 8753B network analyzer. The measurement setup and measured response are illustrated in Figures 3.76 and 3.77, respectively. The characteristics of some of the coaxial cables that can be used in the construction of the balun are given in Table 3.4.

The measured values showing the frequency response of the input balun between 1 and 30 MHz are given below.

Frequency (MHz)	Real	Imaginary	Inductance (nH)
1	51.252	0.238	37.923
5	51.393	0.8789	27.976
13.56	51.734	2.1484	25.216
27.12	52.777	4.168	24.44

Design Example: Implementation of Output Balun

Design an output balun for a push–pull amplifier, which has 12.5 Ω load line impedance. The output power from the balun is desired to be 5000 W at the operational frequency, $f = 13.56$ MHz.

SOLUTION

The equivalent circuit of the output balun is given in Figure 3.78.

It is given that the load power is 5000 W and line impedances are 12.5 Ω.

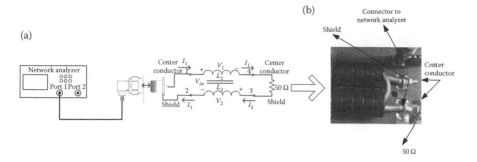

FIGURE 3.76 (a) Block diagram for measurement of the frequency response of the input balun. (b) Measurement setup for frequency response for the input balun.

$$P_L = 5000\,[\text{W}] \quad \text{and} \quad 0.5R = 12.5\,[\Omega]$$

So, the load impedance and power on each pair are found as

$$R = R_L = \frac{12.5}{0.5} = 25\,[\Omega], \quad P = \frac{P_L}{2} = 2500\,[\text{W}]$$

The operational voltage is found as

$$V_{rms} = \sqrt{(2500)12.5} = 176.78\,[\text{V}]$$

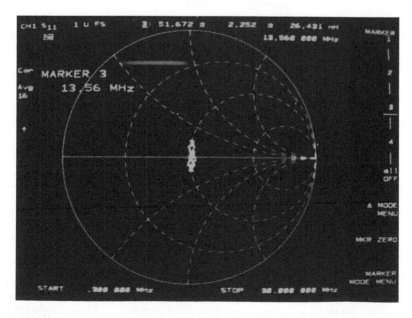

FIGURE 3.77 Frequency response of the input balun.

TABLE 3.4
Coax Cable Information

Coax Cable	OD (in)	Impedance	Jacket Material	Center Conductor Material	Center Conductor Type	Center Conductor Diameter (in)	Dielectric
RG142	0.195	50	Foamed polyethylene	Silver-plated copper steel wire	Solid	0.37	PTFE
RG178	0.071	50	Foamed polyethylene	Silver-plated copper steel wire	Stranded	0.0212	PTFE
RG179	0.1	75	Foamed polyethylene	Silver-plated copper steel wire	Stranded	0.012	PTFE
RG188	0.1	50	PTFFE	Silver-plated copper steel wire	Stranded	0.0201	PTFE
RG316	0.098	50	Foamed polyethylene	Silver plated copper steel wire	Stranded	0.0201	PTFE
RG316	0.114	50	Foamed polyethylene	Silver-plated copper steel wire	Stranded	0.0201	PTFE
RG400	0.2	50	Foamed polyethylene	Silver-plated copper steel wire	Stranded	0.038	PTFE
RG393	0.39	50	Foamed polyethylene	Silver-plated copper steel wire	Stranded	0.094	PTFE
RG58	0.195	50	PVC-NC	Tinned copper	Stranded	0.036	PE
RG59	0.242	75	PVC-NC	Bare copper	Solid	0.023	PE
RG214	0.425	50	PVC-NC	Silver-plated copper	Stranded	0.089	PE
RG393	0.212	50	PVC-NC	Silver-plated copper	Solid	0.036	PE

Then, the load voltage is

$$V_{L,rms} = 2V_{rms} = 2 \times 176.78 = 353.56\,[\text{V}]$$

At this point, we are ready to calculate the required inductance value, which is used to identify the core that will be implemented in the construction of the output balun. The target impedance from the rule of thumb is 5–10 times higher than the higher impedance termination. The target impedance is then

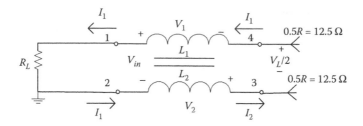

FIGURE 3.78 The equivalent circuit of the output balun.

$$Z_{req} \approx 10(25) = 250 \, [\Omega]$$

The target inductance value is found from

$$Z = 2\pi f L = 2\pi(13.56 * 10^6) * L = 250 \, \Omega$$

which leads to $L \geq 2.934 \, \mu H$. We can now calculate the required number of turns to obtain the target inductance as

$$L = \frac{0.4\pi\mu(\text{relative})A_e \, (\text{cm}^2)N^2}{100 l_e \, (\text{cm})} \, [\mu H]$$

Let us choose the binocular core illustrated in Figure 3.79 for the construction of our balun. The material properties of the core are given in Figure 3.80.

The magnetic path length, cross-sectional area, and thickness of the core is found from the physical dimensions and geometry of the core as $l_e = 8.755$ cm, $A_e = 2.4587$ cm², and $h = 2.54$ cm. Two of these cores are stacked together to increase the saturation flux density of the structure. Hence, the overall cross-sectional area is $A_{eTot} = 2 * A_{eT} = 4.9174$ cm². The required number of turns is then obtained from

$$N = \sqrt{\frac{L \times l_e}{4\pi\mu_i A_{Tc}}} = \sqrt{\frac{2934 \times 8.755}{4\pi \times 40 \times 4.9174}} = 3.22 \sim 3 \text{ turns}$$

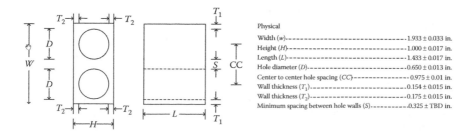

Physical

Width (w)	1.933 ± 0.033 in.
Height (H)	1.000 ± 0.017 in.
Length (L)	1.433 ± 0.017 in.
Hole diameter (D)	0.650 ± 0.013 in.
Center to center hole spacing (CC)	0.975 ± 0.01 in.
Wall thickness (T₁)	0.154 ± 0.015 in.
Wall thickness (T₂)	0.175 ± 0.015 in.
Minimum spacing between hole walls (S)	0.325 ± TBD in.

FIGURE 3.79 The binocular core for the output balun.

Note: T_c = 25°C unlesss otherwise noted.

Characteristic	Symbol	Min.	Typ.	Max.	Unit
Initial permeability (f = 13.56 MHz)	(μ_1)	32	40	40	—
Maximum permeability (f = 50 MHz)	μ_m	—	80	—	—
Saturation flux density ($H = H_{max}$ = 20 oersteds)	B_s	2100	—	—	Gauss
Residual flux density	B_r	—	—	1000	Gauss
Coercive force	H_c	—	—	15	Oersteds
Curie temperature	T_c	450			°C
Temperature coefficient of initial perm. (20°C to 120°C)		—	—	0.1	% per °C
Loss factor (f = 13.56 MHz)	$\dfrac{\text{Tan}}{\mu}$	—	—	80	×10⁻⁶
Volume resistivity	p	1×10^6	—	—	Ω-cm

FIGURE 3.80   Material properties of the output balun core.

The calculated inductance with rounded number of turns is

$$L = \frac{0.4\pi(40)(4.9174)(3^2)}{100(8.755)} \, [\mu H] = 2.54 \, [\mu H]$$

So, the inductance requirement is satisfied since we used a very conservative rule of thumb in calculating that value. As a result, this is the right core configuration with three turns. We need to determine now the wire that will be used for winding. The windings can be done with two different configurations of transmission lines as discussed earlier. One of the configurations is using copper strip configuration. The other configuration is using the twisted wires. The target characteristic impedance for transmission lines is Z_o = 25 Ω. 25 Ω characteristic impedance using copper strip in stripline configuration was calculated earlier. 25 Ω characteristic impedance using twisted lines can be obtained by paralleling four of 2 × 18 AWG wires. The characteristic impedance of 2 × 18 AWG is around 120 Ω. When four of them are paralleled, this combination gives around 30 Ω characteristic impedance, which is acceptable for our core configuration.

The same calculation can be done using the program that is developed for balun design. The design parameters are entered to the program and outputs are identical to the ones that are calculated. The results, including operational current, voltage, flux, inductance value, and required number of turns, are displayed as the output of MATLAB GUI as shown in Figure 3.81. The output balun is constructed as shown in Figure 3.82. It is connected to HP 4191A RF impedance analyzer and its inductance value is measured at 13.56 MHz. The measured inductance value is 3.06 μH as shown in Figure 3.83 and is close to the calculated value at 13.56 MHz.

The output balun is then connected to the network analyzer, HP 8753B, and its frequency response between 1 and 30 MHz is illustrated in Figure 3.84b.

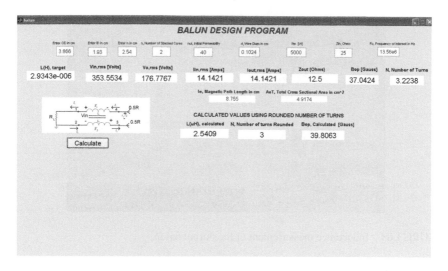

FIGURE 3.81 Output balun design program.

FIGURE 3.82 Constructed output balun.

The measurement setup is illustrated in Figure 3.84a.The input of the balun is terminated with 25 Ω impedance and the output is connected to the network analyzer and one-port measurement has been done as illustrated in Figure 3.84. The measured values shown in Figure 3.85 illustrates the frequency response of the output balun between 1 and 13.56 MHz and are given as

Frequency (MHz)	Real	Imaginary	Inductance (nH)
1	25.012	0.75	119.3
5	25.07	3.39	107.9
13.56	28.21	10.01	117.37

FIGURE 3.83 Inductance measurement of the output balun.

(b)

(a)

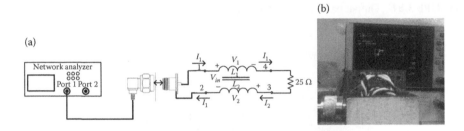

FIGURE 3.84 (a) Block diagram for measurement of the balun frequency response. (b) Measurement of the frequency response.

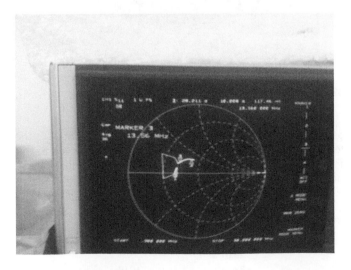

FIGURE 3.85 Frequency response of the output balun.

It is important to note that the frequency response of the output balun is better at lower frequencies than higher frequencies. The frequency response can be improved by using the compensation techniques with lumped elements discussed earlier.

Design Example: Combination of Balun and TLTs

Design a passive feeding system for a push–pull amplifier from 50 Ω source with 10 W power to feed 2.78 Ω load line at 27.12 MHz. Accomplish this by using the input balun and TLTs. Build balun and TLTs, simulate, measure, and compare the results.

SOLUTION

The design practice is to use the minimum number of parts to accomplish the task. It is obvious that the input balun will divide the signal equally with 180° phase shift to feed the lines of the push–pull amplifier. The output of the input balun on each line is 25 Ω. We can then use TLT to transform 25 Ω to 2.78 Ω to feed the push–pull load lines. This requires the use of 9:1 impedance ratio TLTs. The feeding system for push–pull amplifier can be implemented in two different ways.

i. The first way is to use an input balun and two identical 3:1 voltage ratio or 9:1 impedance ratio TLTs as illustrated in Figure 3.86. Since the source impedance is given to be $R = 50\,\Omega$, then the output of the balun on each line is $0.5R = 25\,\Omega$. When the output of the balun is interfaced with 3:1 voltage ratio TLT, impedance at its output will be $0.0556R = 2.78\,\Omega$. The current at the output of TLT is then 3 times higher than the source current and is equal to $3I$. The voltage at the output of TLT is 6 times less than V_{in}. For maximum power transfer with 10 W input power, the voltage at the input of the balun is

$$V_{in,rms} = \sqrt{P \times R} = \sqrt{10 \times 50} = 22.36\,[\text{V}]$$

This requires the source voltage and the current to be

$$V_{s,rms} = 2(V_{in,rms}) = 44.72\,[\text{V}], \quad I_{rms} = \frac{V_{in,rms}}{50} = 0.4472\,[\text{A}]$$

The current and voltage at the output of the balun and input of the TLT are

$$V_{rms} = \frac{V_{in,rms}}{2} = 11.18\,[\text{V}], \quad I_{rms} = 0.4472\,[\text{A}]$$

The current and voltage at the output of TLT are calculated to be

$$V_{rms} = \frac{V_{in,rms}}{6} = 3.726\,[\text{V}], \quad I_{TLT,rms} = 3I_{rms} = 1.3416\,[\text{A}]$$

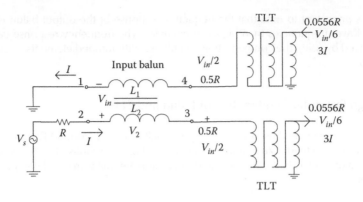

FIGURE 3.86 The first way of implementing the feeding system.

Pspice simulation circuit of the first feeding system is shown in Figure 3.87 and simulation results for the voltage and current waveforms are shown in Figures 3.88 and 3.89, respectively. The calculated values for the voltages and currents are the rms values whereas the values shown in Figures 3.88 and 3.89 illustrate the peak values.

As can be seen from Figures 3.88 and 3.89, the simulated and calculated values are in close agreement.

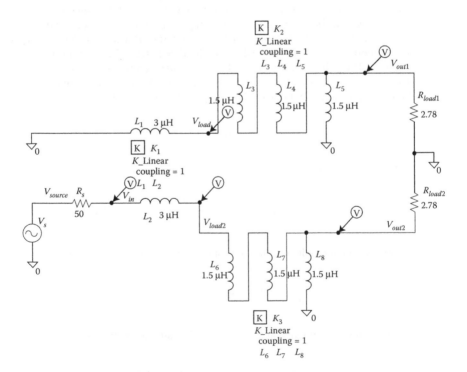

FIGURE 3.87 Pspice simulation of the first feeding system.

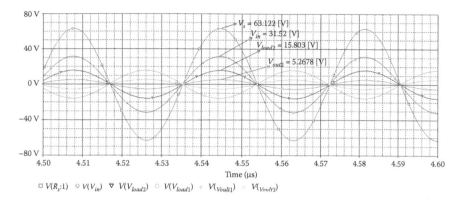

FIGURE 3.88 Pspice simulation results for the voltage waveforms.

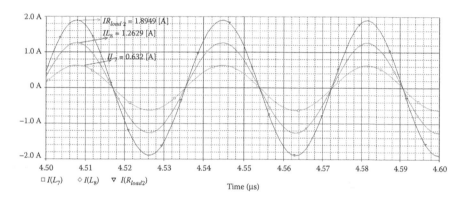

FIGURE 3.89 Pspice simulation results for the current waveforms.

ii. The second way of implementation of TLT utilizes hybrid connection, which is illustrated in Figure 3.90. The benefit of hybrid connection among TLTs is that it reduces the number of transmission lines used in the construction of TLT.

The amount of current and voltage on each line is illustrated in Figure 3.90. The feeding system shown in Figure 3.90 is simulated using Pspice as shown in Figure 3.91.

The peak values of the currents and voltages shown in Figures 3.92 and 3.93 are identical to the ones obtained in Figures 3.88 and 3.89. Hence, two methods result in the same frequency response and as a result it is more advantageous to implement TLT using the second method since the number of transmission lines are reduced in comparison to the TLT implemented with the first method.

The implementation of (i) input balun and (ii) TLT used in the feeding system are described in the following section.

iii. *Implementation of input balun*   The target impedance value for the input balun is set to be 10 times higher than the highest impedance termination as we discussed earlier. It is then

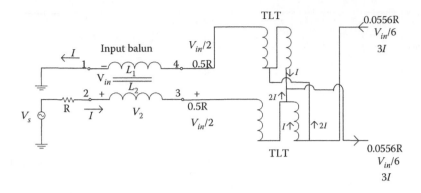

FIGURE 3.90 Hybrid connection of the feeding system for the push–pull amplifier.

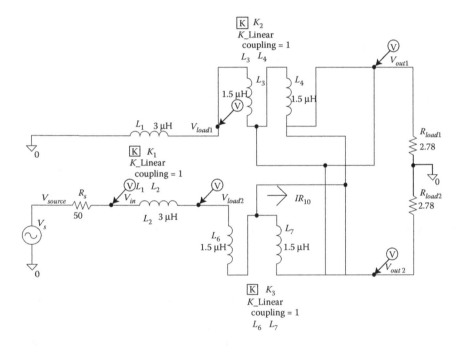

FIGURE 3.91 Pspice simulation of the second feeding system.

$$Z_{req} = 2\pi f L = 2\pi(27.12 \times 10^6) \times L = 500\,\Omega$$

Then

$$L \geq \frac{500}{2\pi(27.12 \times 10^6)} \quad \text{or} \quad L \geq 2.934\,\mu H$$

The number of turns is found from the inductance equation given below

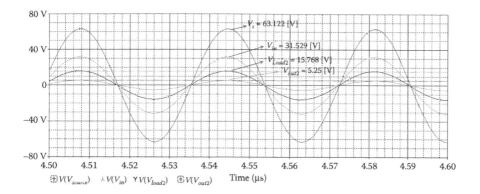

FIGURE 3.92 Pspice simulation results for the voltage waveforms.

$$L = \frac{4\pi N^2 \mu_i A_{TC}}{l_e} \text{ [nH]}$$

We choose Ferronics toroid core with manufacturer part number 11-270. This is a J-type material with a permeability of 850 at 0.1 MHz, $\mu_i = 850$, OD = 1.588 cm, ID = 0.89 cm, and $h = 0.47$ cm. The saturation flux density at 0.1 MHz from Table 2.9 is given as 2800 [G]. The geometry of the core is given in Figure 3.94.

Since the operational frequency is 27.12 MHz, the initial permeability and the saturation flux density will be significantly different than what has been specified at 0.1 MHz. The initial permeability can be estimated using the manufacturer measured data for the inductor, which is constructed using J-type core. This data showing inductance and impedance values are shown in Figure 3.95.

As seen from Figure 3.95, the inductance ratio at 1–27 MHz is around

$$\text{Inductance ratio} = \frac{0.29 \text{ [}\mu\text{H]}}{0.05 \text{ [}\mu\text{H]}} = 5.8$$

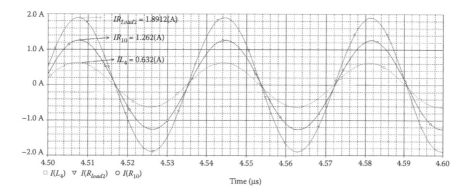

FIGURE 3.93 Pspice simulation results for the current waveforms.

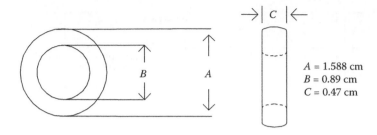

$A = 1.588$ cm
$B = 0.89$ cm
$C = 0.47$ cm

FIGURE 3.94 Geometry of the input balun core.

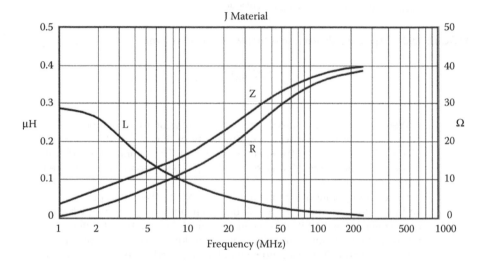

FIGURE 3.95 Manufacturer measured data for the test inductor with J-type core.

FIGURE 3.96 Input balun design program output with the core dimensions.

The inductance difference is mainly due to the change in the permeability. Since the initial permeability that is measured at 0.1 MHz is given to be $\mu_i = 850$, then the permeability of the same core at 27.12 MHz can be approximated to $\mu_i = 850/(5.8) = 146.55$, which can be rounded to 150. This value is used in our input balun design program and the results are illustrated in Figure 3.96. From the program output, it

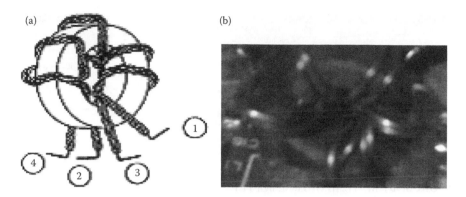

FIGURE 3.97 (a) Construction diagram of the input balun. (b) Implemented input balun design.

is seen that the required number of turns to obtain the target inductance value is six. Winding is done using bifilar TL, which is constructed with 20 AWG enameled wire. The constructed core and its diagram are given in Figure 3.97.

The constructed input balun is measured with HP 4191A RF impedance analyzer at 27.12 MHz. Its inductance value is measured to $L = 2.61$ [µH]. This value is close to the calculated value and the design value obtained by MATLAB GUI.

iv. *Implementation of TLT* Similar procedure is applied to construct TLT for the feeding system. In the construction, we choose Ferronics core with manufacturer part number 11-260. This is again a J-type material with the same properties and initial permeability characteristics discussed earlier. The geometry of the core is given in Figure 3.94. Its physical dimensions are OD = 1.27 cm, ID = 0.714 cm, and h = 0.478 cm. The target impedance is considered to be 10 times higher than the highest termination impedance, which is 25 Ω in this case. Then, the target inductance is found from

$$Z_{req} = 2\pi f L = 2\pi(27.12 \times 10^6) \times L = 250 \ \Omega$$

Then,

$$L \geq \frac{250}{2\pi(27.12 \times 10^6)} \quad \text{or} \quad L \geq 1.4717 \, \mu H$$

The number of turns is found from

$$N = \sqrt{\frac{L \times l_e}{4\pi\mu_i A_{Tc}}} = \sqrt{\frac{1471.7 \times 3.033}{4\pi \times 150 \times 0.133}} = 4.21 \sim 4 \text{ turns}$$

MATLAB GUI has been developed to design TLT based on critical input parameters. The results of GUI are illustrated in Figure 3.98.

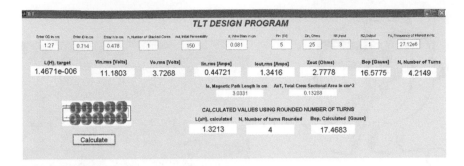

FIGURE 3.98 MATLAB GUI to design TLT.

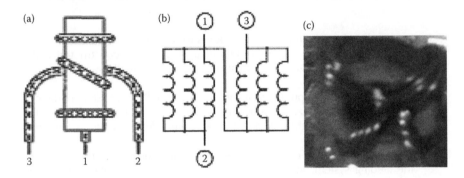

FIGURE 3.99 (a) Construction diagram of the input balun. (b) Circuit diagram of implemented TLT. (c) Implemented input balun design.

The GUI output values are in alignment with the calculated values as expected. When this is placed back into the inductance formula, we obtain the inductance as $L = 1.32$ [μH]. Hence, the design seems to be satisfying the design constraints and can be implemented. The constructed TLT and its construction diagram are illustrated in Figure 3.99. The winding is done using trifilar TL with 22 AWG wire as shown in the figure.

REFERENCES

1. G. Guanella. *New Method of Impedance Matching in Radio-Frequency Circuits*, The Brown Boveri Review, pp. 327–329, September 1944.
2. C.L. Ruthroff. Some broadband transformers, *Proceedings of the IRE*, 47(8), pp. 1337–1342, 1959.
3. O. Pitzalis and T.P.M. Couse. Practical design information for broadband transmission line transformers, *Proceedings of the IEEE Letters*, 56(5), 738–739, 1968.
4. R. Lee, L. Wilson, and C.E. Carter. *Electronic Transformers and Circuits*, 3rd ed., Wiley-Interscience, ISBN 0-471-81976-X, 1988, New York, NY.
5. J. Sevick. *Transmission Line Transformers*, 4th ed., Noble Publishing, 2001, USA.

6. A. Malinen, K. Stadius, and K. Halonen. Characteristics and modeling of a broadband transmission-line transformer, *IEEE 2004 International Symposium on Circuits and Systems*, 4, 413–416, 2004.
7. P. Lefferson. Twisted magnet wire transmission line, *IEEE Transactions on Parts, Hybrids, and Packaging*, 7(4), 148–154, 1971.
8. E. Rotholz. Transmission-line transformers, *IEEE Transactions on Microwave Theory and Techniques*, 29(4), 327–331, 1981.
9. G.A. Breed. Transmission line transformer basics, *Applied Microwave and Wireless*, 10(4), 60, 1998.
10. J. Horn and G. Boeck. Design and modeling of transmission line transformers, *IEEE 2003 International Symposium on Microwave Theory and Techniques*, pp. 421–424, 2003, Philadelphia, US.
11. J.F. Cline. Radio-frequency autotransformers with coaxial-transmission-line windings, *IEEE Transactions on Component Parts*, 119–123, 1961.
12. B. Cogitore, J.P. Keradec, and J. Barbaroux. The two-winding transformer: An experimental method to obtain a wide frequency range equivalent circuit, *IEEE Transactions on Instrumentation and Measurement*, 43(2), 364–371, 1994.

4 MF-UHF Combiner, Divider, and Phase Inverter Design Techniques

4.1 INTRODUCTION

Power combiners, dividers (splitters), and phase inverters have widespread use in RF applications. Power combiners and splitters are passive devices where signals need to be combined or splitted with required insertion loss, good amplitude, and phase balance. These devices are used in applications, including RF/microwave amplifiers, transmitters and receivers, and antenna-array feed networks [1–12].

The concept of combiner is well explained by Wilkinson. The same concept can be applied in the design of splitters too. Wilkinson described a circularly symmetric N-way hybrid power divider having an excellent isolation between individual ports. Wilkinson's power splitter/combiner concept can be realized using lumped elements or distributed elements. Lumped elements involve inductors, which are constructed using toroids, air cores, and so on, that require hand winding as described in Chapter 2. Using these types of elements requires tight control of the parameters such as winding spacing and quality control of magnetics and wire, and increases cost and prevents repeatability from assembly to assembly. The implementation of distributed elements is challenging due to their increased physical length for low operational frequency range. Hence, it is a common practice to use transmission lines' equivalent lumped-element models for applications at low frequency ranges.

4.2 ANALYSIS OF COMBINERS AND DIVIDERS

The $n + 1$ power combiner consists of 1 input port and n output ports as shown in Figure 4.1a, whereas $n + 1$ power divider has n input ports and 1 output port as shown in Figure 4.1b. The analysis of power dividers using the circuit in Figure 4.2 is given by Wilkinson.

The input port is defined as a node where the input resistor R_a is connected. The voltage at node a is defined as V_a. Ports 1 through n are defined as the output ports and voltages at these nodes are designated as V_n. Each transmission line with a characteristic impedance of Z_o and length $l = \lambda/4$ is connected to a common node designated by n via resistor R_x. This can be illustrated in Figure 4.2.

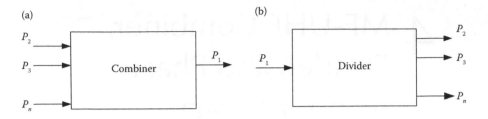

FIGURE 4.1   (a) $n + 1$ port power combiner. (b) $n + 1$ power divider.

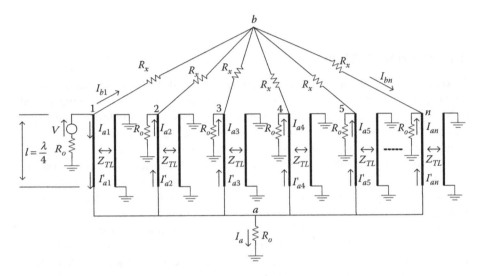

FIGURE 4.2   Equivalent Wilkinson power divider circuit [1].

Currents out of output ports are defined as I_{an} and currents going into input ports are defined as I'_{an}. The transmission lines are basically a quarter of a wavelength away from the input port as illustrated in Figures 4.2 and 4.3. The voltage and current at any point on the transmission lines can be calculated from

$$V(z) = V_o^+ e^{-\gamma z} + V_o^- e^{+\gamma z} \tag{4.1}$$

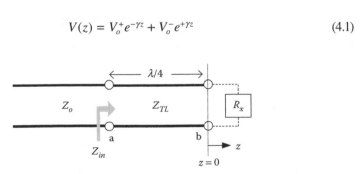

FIGURE 4.3   Quarter wave transmission line connection in the combiner circuit.

$$I(z) = \frac{V_o^+}{Z_o} e^{-\gamma z} - \frac{V_o^-}{Z_o} e^{+\gamma z} \qquad (4.2)$$

The forward and reflected waves are related through reflection coefficient with the following equations:

$$\Gamma = \frac{V_o^-}{V_o^+} = \frac{Z_L - Z_o}{Z_L + Z_o} \qquad (4.3)$$

Assuming the transmission line is lossless, (4.1) and (4.2) can be rewritten in terms of the reflection coefficient as

$$V(z) = V_o^+ (e^{-\beta z} + \Gamma e^{+\beta z}) \qquad (4.4)$$

$$I(z) = \frac{V_o^+}{Z_o} (e^{-\beta z} - \Gamma e^{+\beta z}) \quad \text{or} \quad I(z) = I_o^+ (e^{-\beta z} - \Gamma e^{+\beta z}) \qquad (4.5)$$

At node a, the input voltage can then be expressed using (4.4) as

$$V(z = -l) = V_1 = V_a \left(e^{\frac{\pi}{2}} + \Gamma e^{-\frac{\pi}{2}} \right) \qquad (4.6)$$

Since

$$Z_{TL} = Z_L = Z_o = R_o, \quad \text{then } \Gamma = 0 \qquad (4.7)$$

when (4.7) is substituted into (4.6), it reduces to

$$V_1 = jV_a \qquad (4.8)$$

The current flowing out of node n can be expressed from (4.5) as

$$I(z = -l) = I'_{an} = I_{an} \left(e^{\frac{\pi}{2}} - \Gamma e^{-\frac{\pi}{2}} \right) \qquad (4.9)$$

When (4.7) is substituted into (4.9), it reduces to

$$I'_{an} = jI_{an} \qquad (4.10)$$

The node voltages at nodes n and a are also equal to

$$V_n = I_{an} Z_{TL} \qquad (4.11)$$

$$V_a = I'_{an} Z_{TL} \tag{4.12}$$

When Equation 4.11 is substituted into Equation 4.8, we obtain the voltage at port 1

$$V_1 = jV_a = jI'_{a1} Z_{TL} \tag{4.13a}$$

In addition, voltage at port a can be written in terms of the output current as

$$V_a = jI_{an} Z_{TL} \tag{4.13b}$$

Current at node 1 can be expressed applying KCL as

$$I'_{a1} = I_a + I'_{an}(n-1) = \frac{V_a}{R_o} + I'_{an}(n-1) \tag{4.14}$$

Similarly

$$I_{an} = \frac{V_n}{R_o} - I_{bn} = \frac{V_n}{R_o} - \frac{I_{b1}}{(n-1)} \tag{4.15}$$

When Equation 4.15 is substituted into Equation 4.13, we obtain

$$V_a = jV_n = j\left(\frac{V_n}{R_o} - \frac{I_{b1}}{(n-1)}\right) Z_{TL}$$

The voltage at node b can be written as

$$V_b = V_1 - I_{b1} R_x \tag{4.16}$$

or

$$V_b = V_n + I_{bn} R_x \tag{4.17}$$

Equations 4.16 and 4.17 lead to

$$V_1 - I_{b1} R_x = V_n + I_{bn} R_x \quad \text{or} \quad V_1 - V_n = I_{b1} R_x + I_{bn} R_x \tag{4.18}$$

Equation 4.18 can be modified as

$$V_1 - V_n = R_x(I_{b1} + I_{bn}) \tag{4.19}$$

or

$$V_1 - V_n = R_x\left(I_{b1} + \frac{I_{b1}}{n-1}\right) = R_x I_{b1}\left(\frac{n}{n-1}\right) \tag{4.20}$$

Then

$$I_{b1} = \frac{(V_1 - V_n)}{R_x}\left(\frac{n-1}{n}\right) \tag{4.21}$$

From Equation 4.13a, I'_{a1} can be written as

$$I'_{a1} = -\frac{jV_1}{Z_{TL}} \tag{4.22}$$

Substitution of Equation 4.10 into Equation 4.11 gives

$$I'_{an} = j\frac{V_n}{Z_{TL}} \tag{4.23}$$

When Equations 4.22 and 4.23 are substituted into Equation 4.14, we obtain

$$-\frac{jV_1}{Z_{TL}} = \frac{V_a}{R_o} + j\frac{V_n}{Z_{TL}}(n-1) \tag{4.24}$$

Rewriting Equation 4.24 gives

$$V_a + jV_1\frac{R_o}{Z_{TL}} + j(n-1)V_n\frac{R_o}{Z_{TL}} = 0 \tag{4.25}$$

For perfect isolation, $V_n = 0$, hence Equation 4.25 is simplified to

$$V_a = -jV_1\frac{R_o}{Z_{TL}} \tag{4.26}$$

Substitution of Equation 4.15 into Equation 4.13b gives

$$V_a = j\left(\frac{V_n}{R_o} - \frac{I_{b1}}{(n-1)}\right)Z_{TL} \tag{4.27}$$

When Equation 4.21 is substituted into Equation 4.27, Equation 4.27 is modified as

$$V_a = j\left(\frac{V_n}{R_o} - \frac{(V_1 - V_n)}{nR_x}\right)Z_{TL} \tag{4.28}$$

or

$$V_a - jV_n \frac{Z_{TL}}{R_o} + jV_1 \frac{Z_{TL}}{nR_x} - jV_n \frac{Z_{TL}}{nR_x} = 0 \tag{4.29}$$

which can be further simplified to

$$V_a + jV_1 \frac{Z_{TL}}{nR_x} - jV_n \left(\frac{nR_x Z_{TL} + R_o Z_{TL}}{nR_o R_x} \right) = 0 \tag{4.30}$$

For perfect isolation, $V_n = 0$, so Equation 4.30 is simplified to

$$V_a = -jV_1 \frac{Z_{TL}}{nR_x} \tag{4.31}$$

When Equations 4.26 and 4.31 are compared, we obtain the following relation:

$$\frac{R_o}{Z_{TL}} = \frac{Z_{TL}}{nR_x} \tag{4.32}$$

or

$$Z_{TL} = \sqrt{nR_x R_o} \tag{4.33}$$

Since, for a matched condition, $R_x = R_o$, and then Equation 4.33 can be expressed as

$$Z_{TL} = \sqrt{n} R_o \tag{4.34}$$

Equation 4.34 defines the characteristic impedance of the N-way power combiner/divider when source impedance is R_o. We can now determine the input impedance of the system using Figure 4.3 when $z = -l$ from

$$Z_i = Z_{TL} \frac{R_o + jZ_{TL} \tan \beta l}{Z_{TL} + jR_o \tan \beta l} \tag{4.35}$$

When the length of the transmission line is $l = \lambda/4$, Equation 4.35 is simplified to

$$Z_i = \frac{1}{n} Z_{TL} \frac{R_o + jZ_{TL} \tan((2\pi/\lambda)(\lambda/4))}{Z_{TL} + jR_o \tan((2\pi/\lambda)(\lambda/4))} = \frac{Z_{TL}^2}{nR_o} \tag{4.36}$$

Since $Z_{TL} = \sqrt{n}R_o$ from Equation 4.34, Equation 4.36 can be written as

$$Z_i = \frac{Z_{TL}^2}{nR_o} = \frac{nR_o^2}{nR_o} = R_o \tag{4.37}$$

The isolation between one of the output ports with respect to any other port and VSWR at the input of N-way hybrid power combiner/divider is calculated with the application of superposition using even and odd mode analysis. Figure 4.2 is modified based on the results using the formulations (4.1) through (4.37) and redrawn and shown in Figure 4.4.

The characteristic impedances of the transmission lines in Figure 4.4 are equal to $Z_{TL} = Z_o = \sqrt{N}R_o$ as given by Equation 4.34 and is a quarter wavelength long at the center frequency.

The equivalent four-port network for N-way divider circuit is shown in Figure 4.5. The even mode corresponds to an open circuit at symmetry plane when the voltage source $(+V)$ is placed in series at Port 4, whereas the odd mode corresponds to a short circuit when $(-V)$ is placed in series at Port 4. The networks for even and odd modes are illustrated in Figures 4.6 and 4.7, respectively.

The analysis of even mode network in Figure 4.6 gives the even mode impedance as

$$Z_e = Z_{TL} \frac{((N/N - 1)R_o) + jZ_{TL}\tan(\theta)}{Z_{TL} + j((N/N - 1)R_o)\tan(\theta)} \tag{4.38}$$

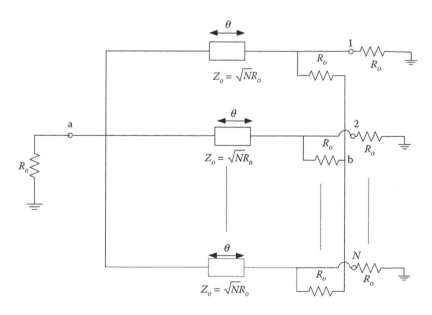

FIGURE 4.4   N-way Wilkinson power divider circuit.

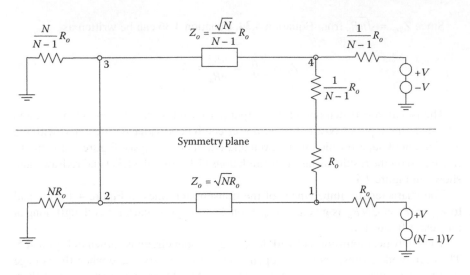

FIGURE 4.5   Four-port network of N-way power divider.

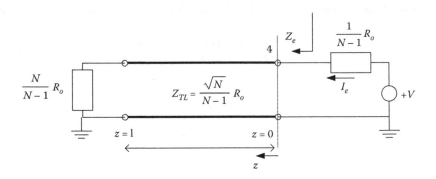

FIGURE 4.6   Even mode network for N-way divider.

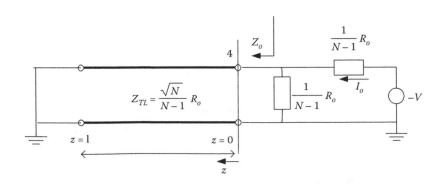

FIGURE 4.7   Odd mode network for N-way divider.

Since

$$Z_{TL} = \frac{\sqrt{N}}{N-1} R_o \tag{4.39}$$

Equation 4.38 can be simplified to

$$
\begin{aligned}
Z_e &= \left(\frac{\sqrt{N}}{N-1} R_o \right) \frac{((N/N-1)R_o) + j((\sqrt{N}/N-1)R_o)\tan(\theta)}{((\sqrt{N}/N-1)R_o) + j((N/N-1)R_o)\tan(\theta)} \\
&= \left(\frac{\sqrt{N}}{N-1} R_o \right) \frac{1 + j(1/\sqrt{N})\tan(\theta)}{(1/\sqrt{N}) + j\tan(\theta)}
\end{aligned}
\tag{4.40}
$$

or

$$Z_e = \left(\frac{\sqrt{N}}{N-1} R_o \right) \frac{\sqrt{N}\cot(\theta) + j}{\cot(\theta) + j\sqrt{N}} \tag{4.41}$$

The even mode network can then be further reduced to the circuit given in Figure 4.8.

$$V = I_e \left(\left(\frac{\sqrt{N}}{N-1} R_o \right) \frac{\sqrt{N}\cot(\theta) + j}{\cot(\theta) + j\sqrt{N}} + \frac{R_o}{N-1} \right) \tag{4.42}$$

as

$$I_e = \frac{V}{R_o} \frac{1}{\left[(\sqrt{N}/N-1)(\sqrt{N}\cot(\theta) + j/\cot(\theta) + j\sqrt{N}) + (1/N-1) \right]} \tag{4.43}$$

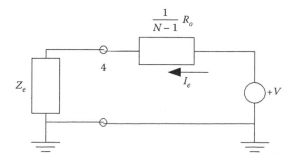

FIGURE 4.8 Simplified even mode network for N-way divider.

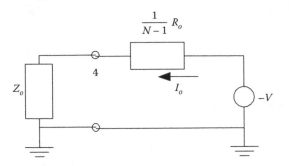

FIGURE 4.9 Simplified odd mode network for N-way divider.

We perform the odd mode analysis similarly and find the odd mode impedance as

$$Z_o = \frac{(R_o/N - 1)(jZ_{TL}\tan(\theta))}{(R_o/N - 1) + (jZ_{TL}\tan(\theta))} \tag{4.44}$$

where Z_{TL} is given by Equation 4.39. The odd mode network can then be further reduced to the circuit given in Figure 4.9.

The odd mode current is found from

$$V = -I_o\left(\frac{R_o}{N-1}\right)\left(1 + \frac{jZ_{TL}\tan(\theta)}{(R_o/N - 1) + jZ_{TL}\tan(\theta)}\right) \tag{4.45}$$

as

$$I_o = -V\left(\frac{N-1}{R_o}\right)\left[\frac{\cot(\theta) + j\sqrt{N}}{\cot(\theta) + j2\sqrt{N}}\right] \tag{4.46}$$

The total current now is found by adding the odd and even mode currents as given by Equation 4.47.

$$I_t = I_e + I_o = \frac{V}{R_o}\left[\frac{1}{\left[\left(\sqrt{N}/N - 1\right)\left(\sqrt{N}\cot(\theta) + j/\cot(\theta) + j\sqrt{N}\right) + \left(1/N - 1\right)\right]}\right.$$
$$\left. -(N-1)\left[\frac{\cot(\theta) + j\sqrt{N}}{\cot(\theta) + j2\sqrt{N}}\right]\right] \tag{4.47}$$

The power delivered to each of $(N-1)$ ports with load resistance R_o is then equal to

$$P = |I_t|^2\left(\frac{R_o}{N-1}\right)\left(\frac{1}{N-1}\right) = |I_t|^2\left(\frac{R_o}{(N-1)^2}\right) \tag{4.48}$$

The power that is available from the excitation port is defined as P_a and is given by

$$P_a = \frac{(NV)^2}{4R_o}$$

(4.49)

The isolation between one port to others is defined as

$$\text{Isolation (dB)} = 10\log\left(\frac{P_a}{P}\right)$$

(4.50)

Substitution of Equations 4.48 and 4.49 into Equation 4.50 gives

Isolation (dB) =

$$10\log\left(\frac{N^2}{4\left[\cot(\theta) + j\sqrt{N}/(N+1)\cot(\theta) + j2\sqrt{N}\right] - \left[\cot(\theta) + j\sqrt{N}/\cot(\theta) + j2\sqrt{N}\right]^2}\right)$$

(4.51)

The input VSWR of the system that is calculated at node a for N-way divider is found from

$$\text{VSWR} = \frac{1 + |\Gamma|}{1 - |\Gamma|}$$

(4.52)

where

$$\Gamma = \frac{Z_i - R_o}{Z_i + R_o}$$

(4.53)

$$Z_i = \frac{1}{N}\left(Z_{TL}\frac{R_o + jZ_{TL}\tan\theta}{Z_{TL} + jR_o\tan\theta}\right) = \left(\frac{R_o}{\sqrt{N}}\right)\frac{1 + j\sqrt{N}\tan\theta}{\sqrt{N} + j\tan\theta}$$

(4.54)

The insertion loss at each port is defined as

$$\text{IL (dB)} = 10\log\left(\frac{N}{1 - \Gamma^2}\right)$$

(4.55)

The response of the calculated isolation versus electrical length is given in Figure 4.10. The isolation goes toward infinity when the electrical length is $\theta = 90°$ or $l = \lambda/4$ as expected.

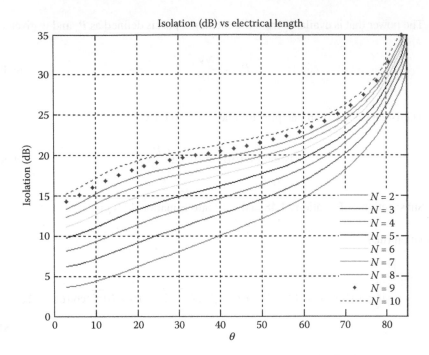

FIGURE 4.10   The isolation response versus electrical length for N-way divider.

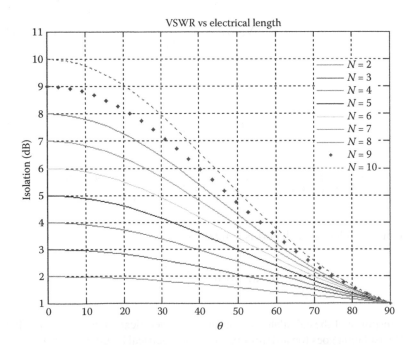

FIGURE 4.11   The VSWR response versus electrical length for N-way divider.

The response of the calculated input VSWR and insertion loss versus electrical length are given in Figures 4.11 and 4.12, respectively. The insertion loss at a quarter wavelength transmission line is tabulated and shown in Table 4.1. The important note on the divider design that is given is that when there is no phase difference between ports, it can also be used as a power combiner by simply applying signals at the output ports that will be added at the single input port.

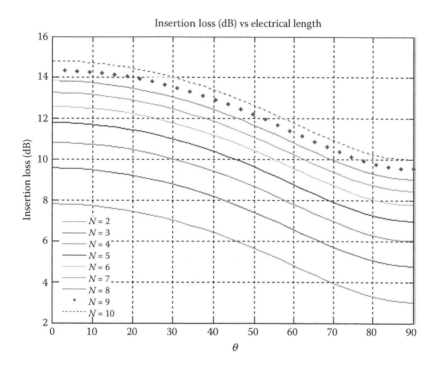

FIGURE 4.12   The insertion loss response versus electrical length for N-way divider.

TABLE 4.1
Calculated Insertion Loss (dB) for $\theta = 90°$

Number of Ports	Calculated Insertion Loss (dB)
2	3.0103
3	4.7712
4	6.0206
5	6.9897
6	7.7815
7	8.4510
8	9.0309
9	9.5424
10	10

4.3   ANALYSIS OF DIVIDERS WITH DIFFERENT SOURCE IMPEDANCE

The analysis that was given in Section 4.2 was based on the assumption that the source and load impedances were equal. However, there might be cases where the source impedance is different from the load impedance as shown in Figure 4.13. When this is the case, the characteristic impedance of the transmission line is then defined from Equation 4.33 as

$$Z_{TL} = \sqrt{nR_xR_o} = \sqrt{nR_gR_o} \tag{4.56}$$

where $R_x = R_g$ is the source impedance.

The equivalent four-port network can be established similarly as shown in Figure 4.14 for even and odd mode analysis based on the application of superposition.

The even and odd mode networks obtained from the circuit given in Figure 4.14 are illustrated in Figures 4.15 and 4.16, respectively.

When a similar analysis in Section 4.2 is performed, the even and odd mode currents are obtained as

$$I_e = \frac{V}{R_o} \frac{1}{\left[\left(\sqrt{N(R_g/R_o)}/N - 1 \right) \left[\sqrt{N(R_g/R_o)} \cot(\theta) + j/\cot(\theta) + j\sqrt{N(R_g/R_o)} \right] + (1/N - 1) \right]} \tag{4.57}$$

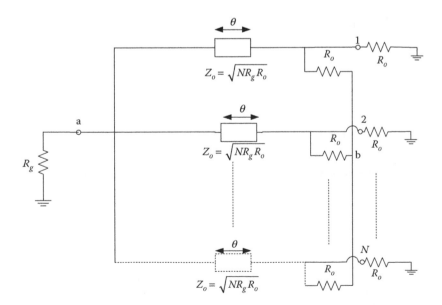

FIGURE 4.13   N-way Wilkinson power divider circuit with different source impedance.

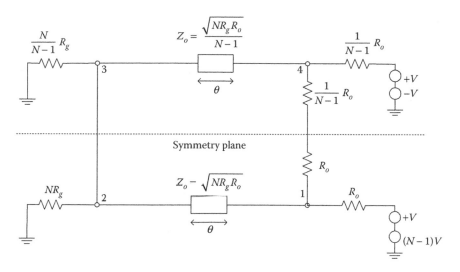

FIGURE 4.14   Four-port network of N-way power divider with different source impedance.

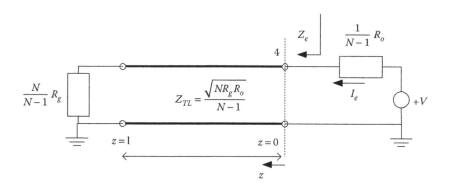

FIGURE 4.15   Even mode network for N-way divider with different source impedance.

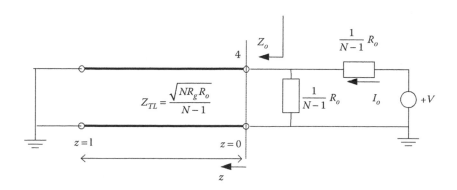

FIGURE 4.16   Odd mode network for N-way divider with different source impedance.

and

$$I_o = -V\left(\frac{N-1}{R_o}\right)\left[\frac{\cot(\theta) + j\sqrt{N(R_g/R_o)}}{\cot(\theta) + j2\sqrt{N(R_g/R_o)}}\right] \tag{4.58}$$

The total current from superposition is then equal to

$$I_t = I_e + I_o = \frac{V}{R_o}\left[\frac{1}{\left[\left(\sqrt{N(R_g/R_o)}/N - 1\right)\left(\begin{array}{c}\sqrt{N(R_g/R_o)}\cot(\theta) + j/\cot(\theta)\\ +j\sqrt{N(R_g/R_o)}\end{array}\right)\right.} \right.$$
$$\left. + (1/N - 1)\right]$$
$$\left. -(N-1)\left[\frac{\cot(\theta) + j\sqrt{N(R_g/R_o)}}{\cot(\theta) + j2\sqrt{N(R_g/R_o)}}\right]\right] \tag{4.59}$$

Isolation and insertion loss are found from Equations 4.50 and 4.55, respectively. The input impedance with different source impedance is given by

$$Z_i = \frac{1}{N}\left(Z_{TL}\frac{R_o + jZ_{TL}\tan\theta}{Z_{TL} + jR_o\tan\theta}\right) = \left(\sqrt{\frac{R_g R_o}{N}}\right)\frac{1 + j\sqrt{N(R_g/R_o)}\tan\theta}{\sqrt{N(R_g/R_o)} + j\tan\theta} \tag{4.60}$$

The reflection coefficient at the input is calculated using the equation

$$\Gamma_{in} = \frac{Z_i - R_g}{Z_i + R_g} \tag{4.61}$$

The input VSWR is then equal to

$$VSWR_{in} = \frac{1 + |\Gamma_{in}|}{1 - |\Gamma_{in}|} \tag{4.62}$$

Available power, P_a, is found from Equation 4.49. The isolation is obtained with the substitution of Equation 4.59 into Equation 4.50 as

Isolation (dB) =

$$
10\log\left(\cfrac{N^2}{4\left|\begin{array}{l}\left[\cot(\theta) + j\sqrt{N(R_g/R_o)}/(N(R_g/R_o)+1)\cot(\theta)\right] \\ +j2\sqrt{N(R_g/R_o)} - \left[\cot(\theta) + j\sqrt{N(R_g/R_o)}/\cot(\theta) + j2\sqrt{N(R_g/R_o)}\right]\end{array}\right|^2} \right) \quad (4.63)
$$

The VSWR at the output ports is calculated using the output reflection coefficient from

$$
\Gamma_o = \left[\cfrac{2\sqrt{N(R_g/R_o)}((R_g/R_o)-1) + j\cot(\theta)(N(R_g/R_o)+1-2(R_g/R_o))}{2\sqrt{N(R_g/R_o)}(2 + N(R_g/R_o)) + j\cot(\theta)(N(R_g/R_o)(4\tan^2(\theta)-1)-1)}\right]
$$

$$(4.64)$$

Then, output VSWR is found by

$$
\text{VSWR}_{out} = \frac{1 + |\Gamma_{out}|}{1 - |\Gamma_{out}|} \quad (4.65)
$$

The response of isolation for several values of (R_g/R_o) is given by Figures 4.17 and 4.18. Figure 4.17 illustrates the isolation versus electrical length when $R_g > R_o$, whereas Figure 4.18 gives the isolation versus electrical length when $R_g < R_o$. It has been observed that the isolation between ports for the N-way power divider with different source impedance gets better when $R_g < R_o$. The input and output VSWR versus electrical length when $R_g > R_o$ and $R_g < R_o$ are given in Figures 4.19 through 4.22. The insertion loss response is given in Figures 4.23 and 4.24 for $R_g > R_o$ and $R_g < R_o$.

It should be noted that the variation in insertion loss versus electrical length is much less when $R_g > R_o$. Hence, it is possible to obtain broadband response when $R_g < R_o$.

Example

Design an eight-way combiner at 100 MHz by assuming (a) $R_g = R_o = 50\ \Omega$ and (b) $R_g = 50[\Omega]$, $R_o = 25[\Omega]$, and find the (i) isolation, (ii) insertion loss, (iii) input VSWR, and (iv) output VSWR when $\theta = 90°$ and $\theta = 70°$ using analytical formulation and compare these values with the simulation results.

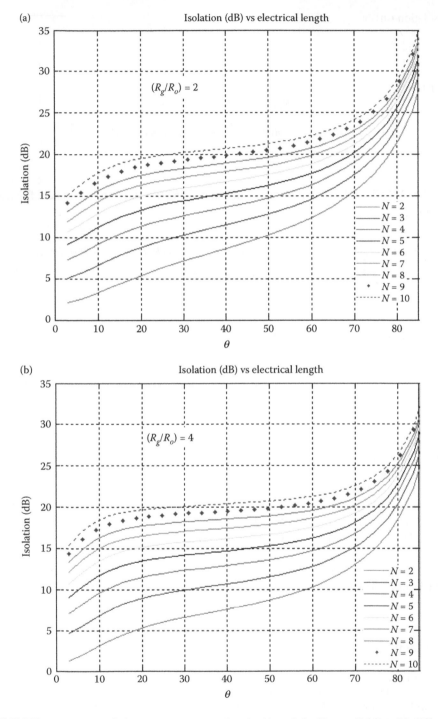

FIGURE 4.17 The isolation response versus electrical length for N-way divider with different source impedance. (a) $(R_g/R_o) = 2$. (b) $(R_g/R_o) = 4$.

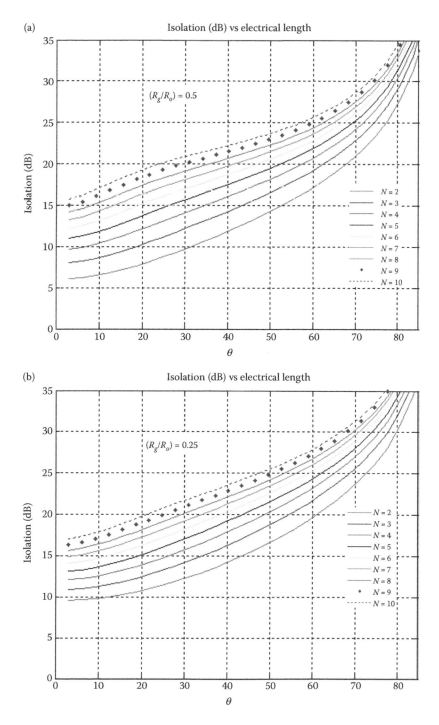

FIGURE 4.18   The isolation response versus electrical length for N-way divider with different source impedance. (a) $(R_g/R_o) = 0.5$. (b) $(R_g/R_o) = 0.25$.

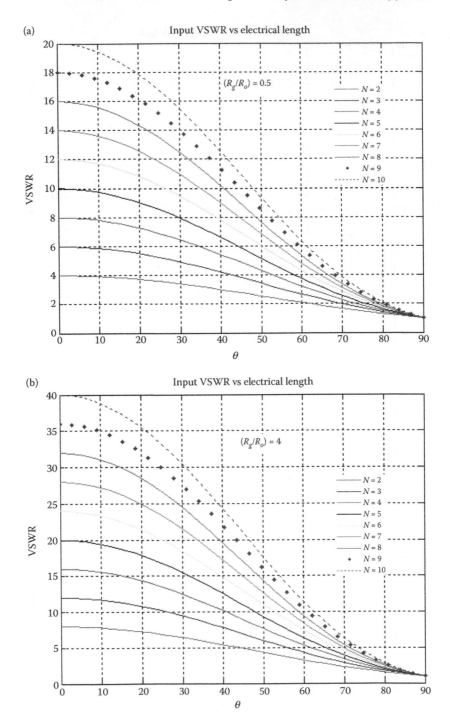

FIGURE 4.19 The input VSWR response versus electrical length for N-way divider with different source impedance. (a) $(R_g/R_o) = 2$. (b) $(R_g/R_o) = 4$.

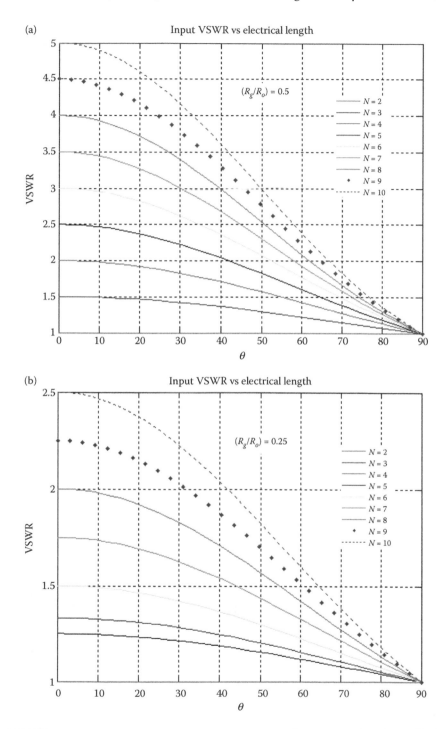

FIGURE 4.20   The input VSWR response versus electrical length for N-way divider with different source impedance. (a) $(R_g/R_o) = 0.5$. (b) $(R_g/R_o) = 0.25$.

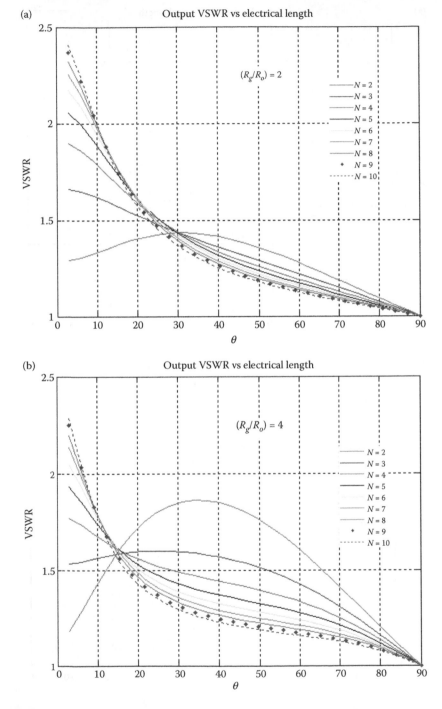

FIGURE 4.21 The output VSWR response versus electrical length for N-way divider with different source impedance. (a) $(R_g/R_o) = 2$. (b) $(R_g/R_o) = 4$.

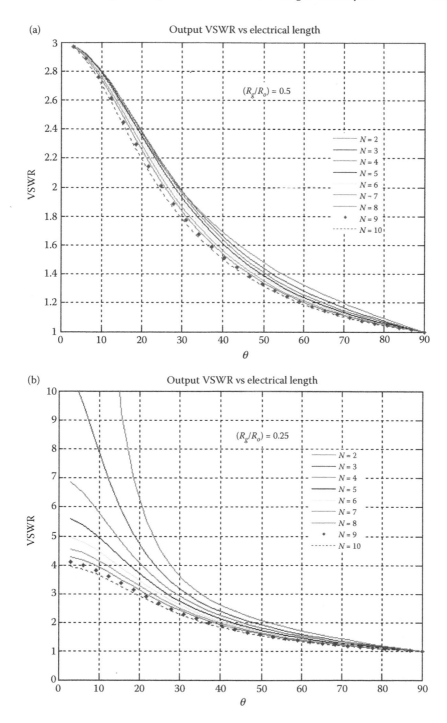

FIGURE 4.22   The output VSWR response versus electrical length for N-way divider with different source impedance. (a) $(R_g/R_o) = 0.5$. (b) $(R_g/R_o) = 0.25$.

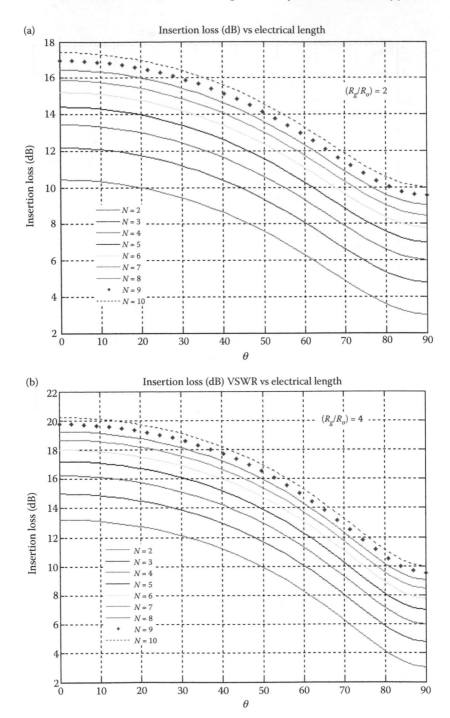

FIGURE 4.23 The insertion loss response versus electrical length for N-way divider with different source impedance. (a) $(R_g/R_o) = 2$. (b) $(R_g/R_o) = 4$.

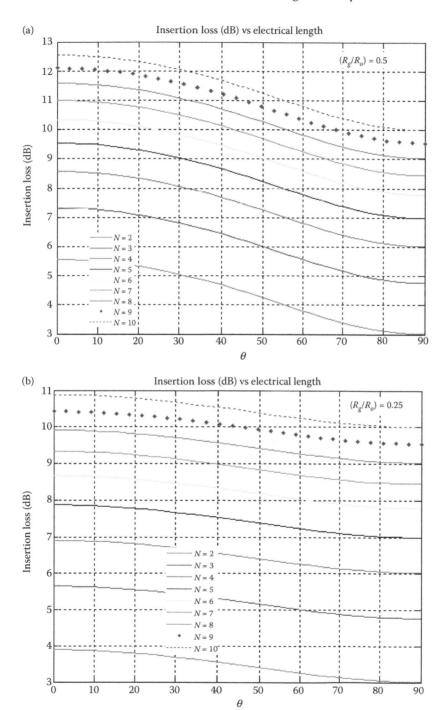

FIGURE 4.24   The insertion loss response versus electrical length for N-way divider with different source impedance. (a) $(R_g/R_o) = 0.5$. (b) $(R_g/R_o) = 0.25$.

SOLUTION

a. In part (a) of the example, a balanced N-way divider is analyzed when $R_g = R_o = 50\ \Omega$.
   i. Isolation for a balanced N-way divider, $R_g = R_o = 50\ \Omega$, is found from (4.51) as

Isolation (dB) =

$$10\log\left(\frac{N^2}{4\left[\cot(\theta) + j\sqrt{N}/(N+1)\cot(\theta) + j2\sqrt{N}\right] - \left[\cot(\theta) + j\sqrt{N}/\cot(\theta) + j2\sqrt{N}\right]^2}\right)$$

Isolation (dB) = 339.3119 at $\theta = 90°$ and isolation (dB) = 25.03 at $\theta = 70°$.
   ii. Insertion loss is calculated from (4.55) as

$$IL\ (dB) = 10\log\left(\frac{N}{1 - \Gamma^2}\right)$$

Insertion loss (dB) = 9.0309 at $\theta = 90°$ and insertion loss (dB) = 9.75 at $\theta = 70°$.
   iii. Input VSWR = 1 at $\theta = 90°$ and input VSWR = 2.27 at $\theta = 70°$.
   The simulation when $R_g = R_o = 50\ \Omega$ has been performed by Ansoft Designer and the results are shown below for each section. The circuit that is simulated is shown in Figure 4.25. In Figure 4.25, port 1 represents the input port and ports 2–9 represent the output ports.
   i. The isolation versus frequency when the electrical length of each transmission line is $\theta = 90°$ is given in Figure 4.26. The isolation is very high as expected when $f = 100$ MHz.
   The isolation versus frequency when electrical length of each transmission line is $\theta = 70°$ is given in Figure 4.27. The isolation is 25.96 dB at $f = 100$ MHz.
   ii. The insertion loss versus frequency when the electrical length of each transmission line is $\theta = 90°$ is given in Figure 4.28. The insertion loss is 9.03 dB when $f = 100$ MHz.
   The insertion loss is 9.57 dB when $f = 100$ MHz for $\theta = 70°$ as shown in Figure 4.29.
   iii. The input and output VSWR versus frequency when $\theta = 90°$ are given in Figure 4.30.
   iv. The output VSWR versus frequency response when $\theta = 70°$ are given in Figure 4.31. As illustrated, all the simulated values are very close to the calculated values.
b. In part (b) of the example, an unbalanced N-way divider is analyzed when $R_g = 50[\Omega]$ and $R_o = 25[\Omega]$
   i. Isolation for a balanced eight-way divider when $R_g = 50[\Omega]$ and $R_o = 25[\Omega]$ is found from (4.63) as

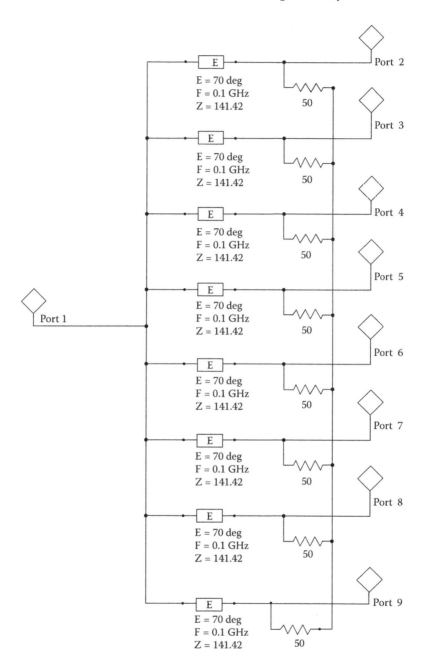

FIGURE 4.25 Simulated eight-way balanced divider.

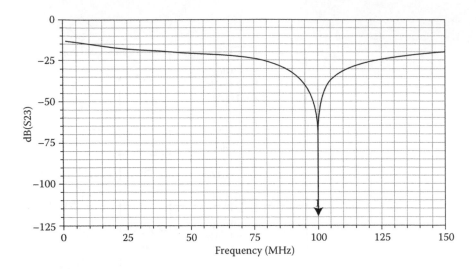

FIGURE 4.26 Isolation versus frequency for eight-way divider when $\theta = 90°$.

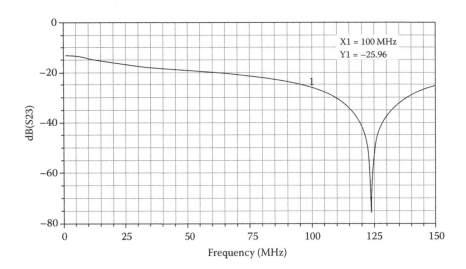

FIGURE 4.27 Isolation versus frequency for eight-way divider when $\theta = 70°$.

$$\text{Isolation (dB)} = 10\log\left(\frac{N^2}{4\left| \begin{array}{l} \left[\cot(\theta) + j\sqrt{N(R_g/R_o)}/(N(R_g/R_o) + 1)\cot(\theta) + j2\sqrt{N(R_g/R_o)}\right]^2 \\ -\left[\cot(\theta) + j\sqrt{N(R_g/R_o)}/\cot(\theta) + j2\sqrt{N(R_g/R_o)}\right] \end{array} \right|} \right)$$

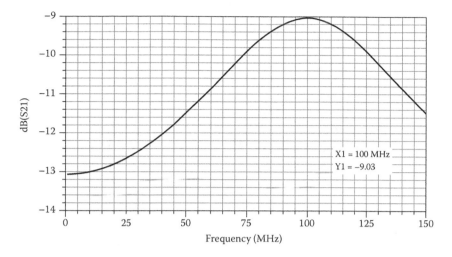

FIGURE 4.28   Insertion loss versus frequency for eight-way divider when $\theta = 90°$.

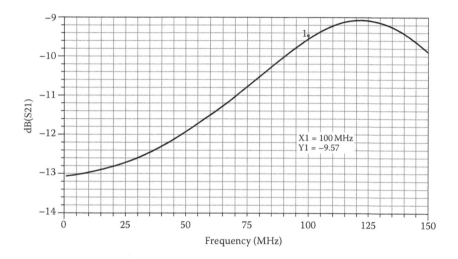

FIGURE 4.29   Insertion loss versus frequency for eight-way divider when $\theta = 70°$.

Isolation (dB) =

$$10\log\left(\frac{8^2}{4\left|\begin{array}{l}\left[\cot(\theta) + j\sqrt{8(50/25)}/(8(50/25) + 1)\cot(\theta) + j2\sqrt{8(50/25)}\right] \\ -\left[\cot(\theta) + j\sqrt{8(50/25)}/\cot(\theta) + j2\sqrt{8(50/25)}\right]\end{array}\right|^2}\right)$$

Isolation (dB) = 336.30 at $\theta = 90°$ and isolation (dB) = 22.8294 at $\theta = 70°$.

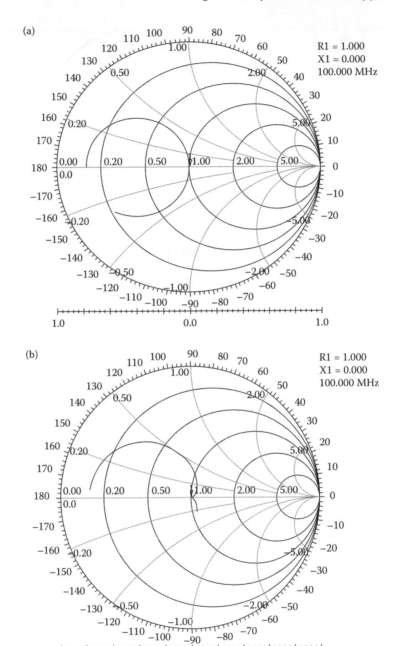

FIGURE 4.30 The input and output VSWR versus frequency when $\theta = 90°$. (a) Input VSWR. (b) Output VSWR.

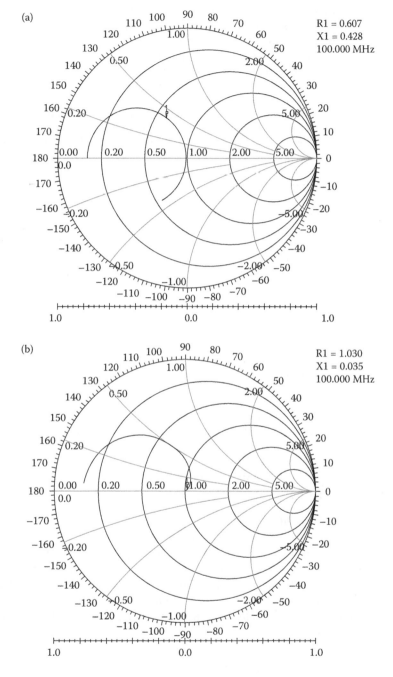

FIGURE 4.31 The input and output VSWR versus frequency when $\theta = 70°$. (a) Input VSWR. (b) Output VSWR.

   ii.   Insertion loss is calculated from (4.55) using (4.64) as IL (dB) = $10\log(N/1 - \Gamma^2)$ Insertion loss (dB) = 9.0309 at $\theta = 90°$ and insertion loss (dB) = 10.5269 at $\theta = 70°$.

   iii. Input VSWR = 1 at $\theta = 90°$ and input VSWR = 3.3461 at $\theta = 70°$.

       The simulation when $R_g = 50[\Omega]$ and $R_o = 25[\Omega]$ has been performed by Ansoft Designer and the results are presented. The circuit that is simulated is shown in Figure 4.32.

       i.   The simulated isolation versus frequency when the electrical length of each transmission line is $\theta = 90°$ and $\theta = 70°$ are given in Figures 4.33 and 4.34, respectively. The simulated isolation values are given in these figures.

       ii.   The simulated insertion loss versus frequency when the electrical length of each transmission line is $\theta = 90°$ and $\theta = 70°$ are given in Figures 4.35 and 4.36, respectively. The simulated insertion loss values are given in these figures.

       iii. The input and output VSWR versus frequency when $\theta = 90°$ and $\theta = 70°$ are given in Figures 4.37 and 4.38, respectively.

       The simulated values are in agreement with the calculated values as shown in part (a) and part (b). This confirms the accuracy of the analytical formulation. As a result, MATLAB® GUI has been developed to design balanced and unbalanced N-way divider using the method presented in Sections 4.2 and 4.3. The program requests source impedance, load impedance, center frequency, number of ports, electrical length and type of divider and it outputs isolation, insertion loss, and input and output VSWRs at the desired electrical length. The program also gives divider characteristic curves, including isolation, insertion loss, and input and output VSWRs versus electrical length. The program output for balanced and unbalanced eight-way divider is given in Figures 4.39 and 4.40, respectively. Results agree with the previously calculated and simulated results.

4.4   MICROSTRIP IMPLEMENTATION OF POWER COMBINERS/DIVIDERS

Consider the N-way combiner circuit given in Figure 4.41. θ in Figure 4.1 is the physical length of transmission line and equal to a quarter of a wavelength, $\lambda/4$. Because of the increased physical length of the distributed components in the HF range, we transform distributed components to lumped-element components. The network showing transformation from distributed elements to lumped elements for a quarter-wavelength-long transmission line is given in Figure 4.42.

In Figure 4.42, the element values for the lumped components can be found using the following formulas:

$$L = \frac{Z_o}{2\pi f}, \quad C = \frac{1}{2\pi f Z_o} \tag{4.66}$$

The network in Figure 4.42 also performs impedance transformation from R_1 to R_2 at each distribution port on the combiner. The lumped-element transformation

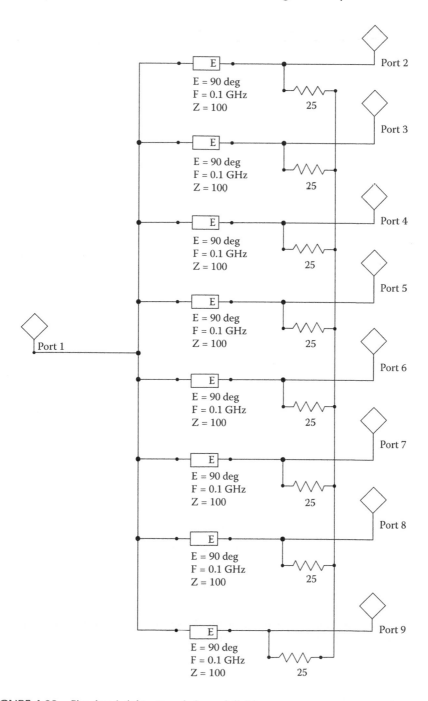

FIGURE 4.32   Simulated eight-way unbalanced divider.

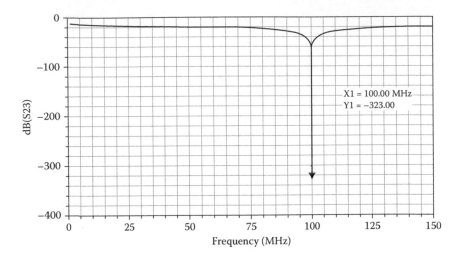

FIGURE 4.33   Isolation versus frequency for eight-way unbalanced divider when $\theta = 90°$.

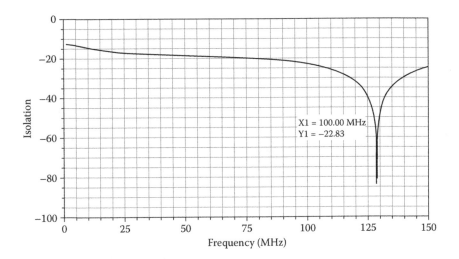

FIGURE 4.34   Isolation versus frequency for eight-way unbalanced divider when $\theta = 70°$.

network shown in Figure 4.43 is a π-network and it consists of three reactive elements. The quality factor for this network can be found from

$$Q = R_1/X_c$$

The number of reactive elements can be reduced by transforming the π-network to the L-network. Q of the π-network can be used to obtain the corresponding element values for the L-network when it is transformed. The equivalent L-network is given in Figure 4.43 when $R_1 \geq R_2$.

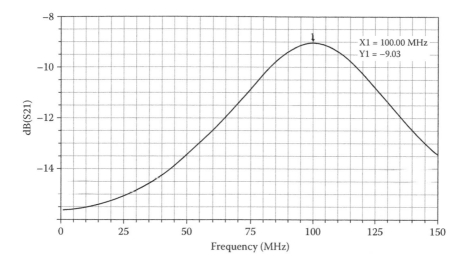

FIGURE 4.35 Insertion loss versus frequency for eight-way unbalanced divider when $\theta = 90°$.

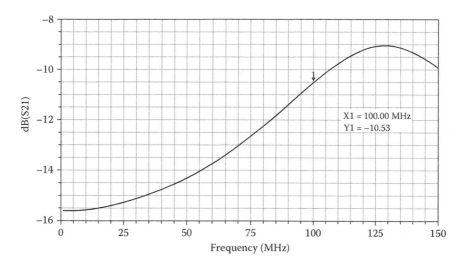

FIGURE 4.36 Insertion loss versus frequency for eight-way unbalanced divider when $\theta = 70°$.

As a result, each distributed element in Figure 4.41 can be replaced with its equivalent lumped-element L-network shown in Figure 4.43b.

Design Example: Three-Way High-Power Microstrip Combiner

Design, simulate, and build and measure a high-power combiner using microstrip technology at 13.56 MHz to combine the output of three PA modules with 25 Ω. The output of the combiner is desired to be 30 Ω.

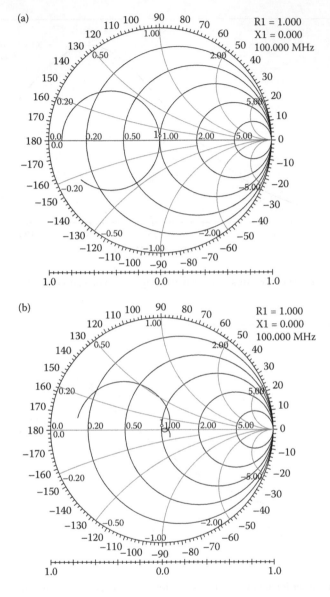

FIGURE 4.37 The input and output VSWR for an unbalanced driver versus frequency when $\theta = 90°$. (a) Input VSWR. (b) Output VSWR.

SOLUTION

The operational frequency is 13.56 MHz, which is a common frequency for ISM (industrial, scientific, and medical) applications, and the combiner should be capable of providing 12,000 W output power. The combiner is intended to combine the outputs of three PA modules. Each PA module presents 25 Ω impedance to the input of each distribution port on the combiner. The output of the combiner

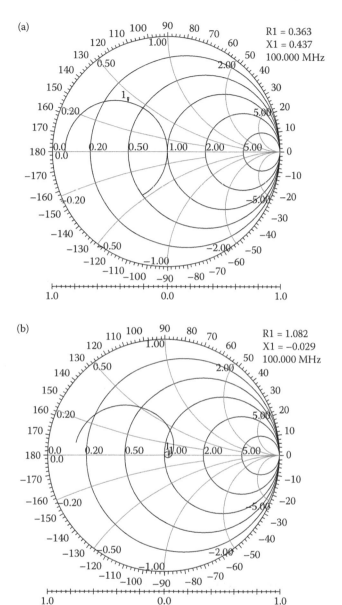

FIGURE 4.38 The input and output VSWR for an unbalanced driver versus frequency when $\theta = 70°$. (a) Input VSWR. (b) Output VSWR.

is desired to be 30 Ω. The analytical formulation developed for dividers can be adapted to combiner analysis. The input parameters are entered into the MATLAB GUI developed for combiners. Insertion loss, isolation, VSWRs, and character-istic curves are obtained and shown in Figure 4.44. The program does not take into account of any imperfections that might exist in the real system and hence

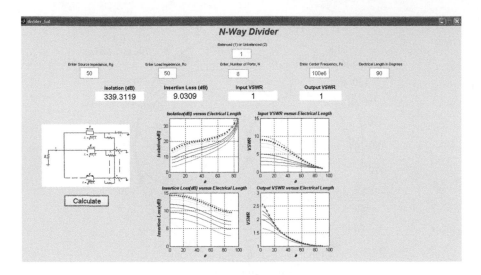

FIGURE 4.39   MATLAB GUI output for eight-way balanced divider when $\theta = 90°$.

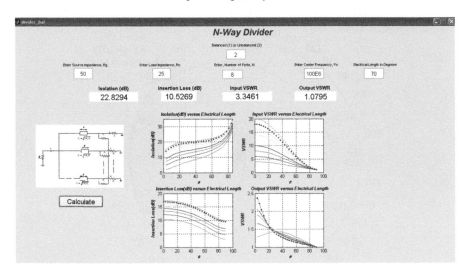

FIGURE 4.40   MATLAB GUI output for eight-way unbalanced divider when $\theta = 70°$.

theoretically gives perfect isolation when $\theta = 90°$. The design parameters for the three-way combiner are

$$R_o = 25\,\Omega, \quad R_L = 30\,\Omega, \quad Z_o = 47.43\,\Omega \tag{4.67}$$

Each PA module is required to provide 4000 W output power under a matched condition. The component values calculated using Equations 4.66 and 4.67 for the π-network given in Figure 4.43a are

$$L_1 = 556.69\,\text{nH}, \quad C_2 = C_3 = 247.46\,\text{pF}, \quad Q = 1.898 \tag{4.68}$$

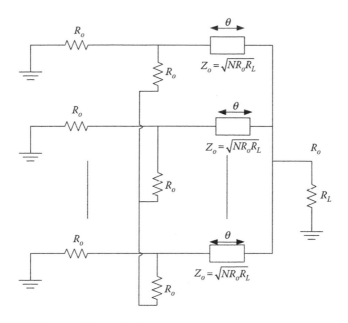

FIGURE 4.41 N-way combiner circuit.

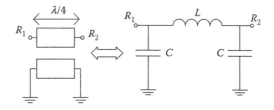

FIGURE 4.42 Distributed to lumped conversion.

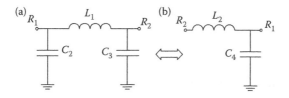

FIGURE 4.43 Transformation from π-network (a) to L-network (b).

The corresponding component values of the lumped elements for the L-network given in Figure 4.43b are

$$L_3 = 472.2\,\text{nH}, \quad C_4 = 210.5\,\text{pF}, \quad Q = 1.612 \tag{4.69}$$

In both circuits

$$R_1 = 90\,\Omega \quad \text{and} \quad R_2 = 25\,\Omega \tag{4.70}$$

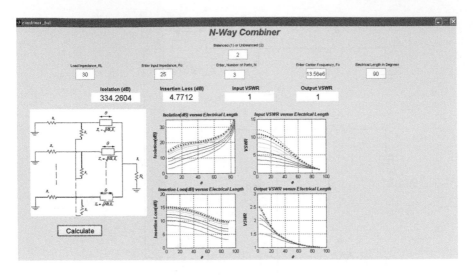

FIGURE 4.44 MATLAB GUI output for three-way unbalanced combiner when $\theta = 90°$.

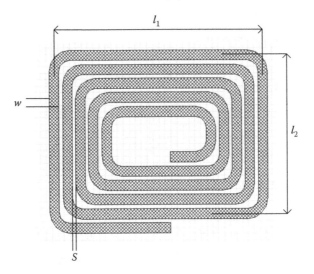

FIGURE 4.45 Spiral inductor layout.

The lumped-element inductor in the L-network is implemented as a spiral inductor on alumina substrate having a planar form. The form of the spiral inductor that will be used in our application is shown in Figure 4.45. It is a rectangular spiral inductor with rounded edges versus sharp edges. This type of implementation on the edges increases the effective arcing distance between traces. The physical dimensions for the spiral inductor are the width of the trace, w, the length of the outside edges, l_1 l_2, and the spacing between the traces, s. The simplified two-port lumped-element equivalent circuit for the spiral inductor shown in Figure 4.45 is illustrated in Figure 4.46. In Figure 4.46, L is the series inductance of the spiral and C is the substrate capacitance. This model ignores the losses in the substrate and

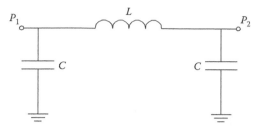

FIGURE 4.46   Simplified equivalent circuit for spiral inductor without loss factor.

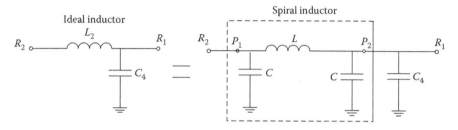

FIGURE 4.47   Spiral inductor model is inserted into L-network.

the conductor. When the two-port equivalent models is inserted into Figure 4.43b to represent the lumped-element model for the spiral inductor, the overall network that represents each distribution circuit can be shown in Figure 4.47. In the proposed design method, first, the lumped-element values of the spiral inductor have to be obtained, and then, physical dimensions of the spiral inductor, which give the desired series inductance, L, are calculated. The effective value of the inductance, L_{eff} of the spiral inductor is measured as a one-port network as shown in Figure 4.48a. The lumped-element values for the spiral inductor to perform the

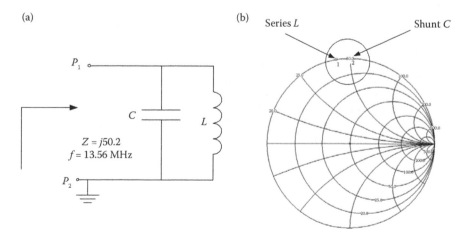

FIGURE 4.48   (a) One-port measurement network and (b) its impedance plot for spiral inductor.

required impedance transformation from $R_2 = 25\,\Omega$ to $R_1 = 90\,\Omega$ for the network in Figure 4.46 are calculated to be

$$L = 497\ \text{nH}, \quad C = 43.6\ \text{pF}$$

Figure 4.48b depicts the impedance plot of the spiral inductor on the Smith chart. The impedance at point 1, P_1, on the Smith chart is purely inductive and equal to $Z = j50.2\,\Omega$. So, the effective value of the inductance for the spiral at the operational frequency using the one-port measurement network in Figure 4.48a is found using $Z = j\,\omega\,L_{eff}$ as

$$L_{eff} = 589.2\ \text{nH} \tag{4.71}$$

The shunt capacitance C_4 in Figure 4.47 will be referred to as C_4' in the existence of the spiral inductor to distinguish it from the value of the model in Figure 4.43b. When the spiral inductor model is used as shown in Figure 4.46, C_4' is calculated accordingly as

$$C_4' = (210.5 - 43.6) = 166.9\ \text{pF} \tag{4.72}$$

The values of the lumped-element components for the distribution circuit in Figure 4.47 using the spiral inductor model are illustrated in Figure 4.49. This circuit represents the final form of the lumped-element L-network distribution circuit that is used as a reference for electromagnetic simulation.

The simulation results shown in Figures 4.50 and 4.51 compare the responses of the π-network and the L-network using ideal lumped elements with circuit simulator. In this figure, the insertion loss of the combiner and the isolation between each distribution port with lumped elements are illustrated using Ansoft Designer. The lumped-element values obtained in Equations 4.66 through 4.67 are used in the simulation. On the basis of the results, the insertion loss is not affected by the transformation of networks from π to L. Although the insertion loss for each network is found to be the same using both topologies, the isolation for each distribution port is much better using π-network at the operational frequency. So, for applications where good isolation between the distribution ports is required, π-network should be employed.

Three-way planar high-power combiner is simulated with the method of moment field solver, Ansoft Designer, using the initial lumped-element values. The spiral inductors are implemented in planar form on a 100-mil-thick Al_2O_3 substrate as detailed in Chapter 2. The planar three-way combiner has three spiral inductors, lumped capacitors, and planar circuitry to combine the structure. The spiral inductor has a bridge-type connection from the excitation point to carry the input signal to the spiral. The simulated structure is shown in Figure 4.52. The

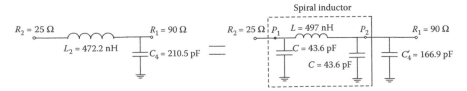

FIGURE 4.49 Final form of the lumped-element distribution circuit with spiral inductor.

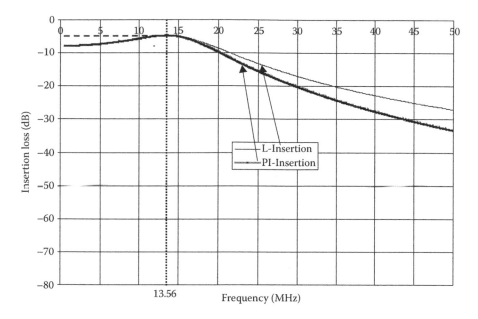

FIGURE 4.50   Simulation results for the insertion loss between each distribution port using π- and L-network.

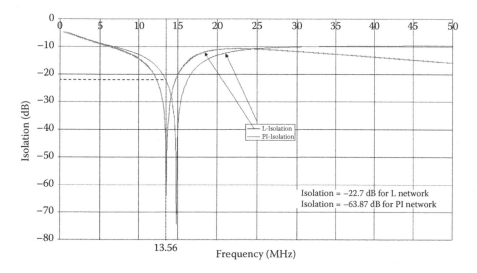

FIGURE 4.51   Simulation results for the insertion loss and isolation between each distribution port using π- and L-network.

simulation results for the whole combiner are shown in Figure 4.53. The current and near field distribution for this structure at the center operational frequency, 13.56 MHz, is given in Figure 4.54. The simulation has been performed by importing the electromagnetic model into the circuit simulator and cosimulating both designs as shown in Figure 4.55. The use of the electromagnetic simulator enables

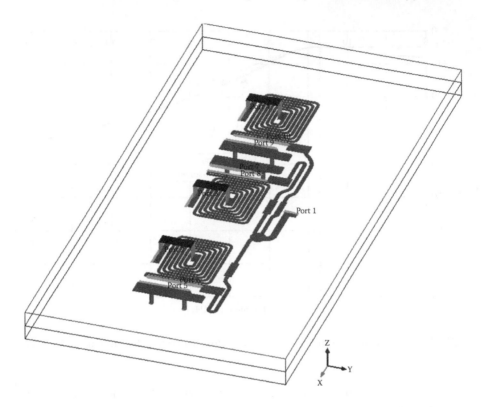

FIGURE 4.52   Simulated three-way combiner in planar form using *L*-network topology.

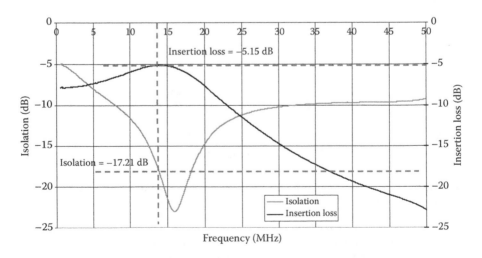

FIGURE 4.53   Simulation results for three-way combiner in planar form using *L*-network topology.

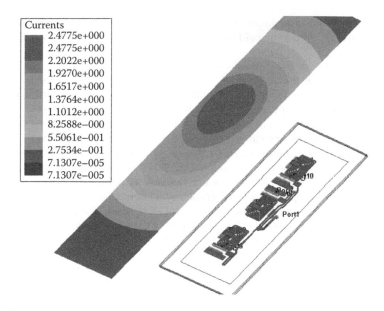

FIGURE 4.54   Current and near field distribution of three-way combiner.

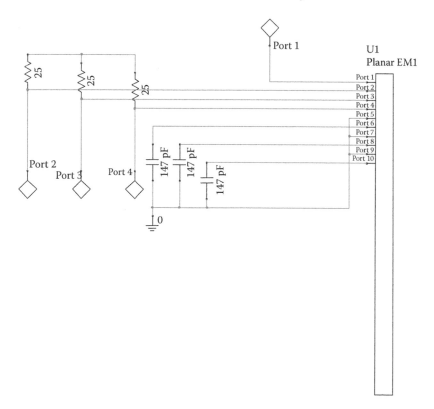

FIGURE 4.55   Cosimulation of three-way combiner.

the designer to observe the electromagnetic coupling and high current distribution regions that need to be paid attention during the implementation stage of the physical design.

On the basis of the simulation results, the insertion loss and the isolation between each port are found to be −5.15 and −17.21 dB, respectively, at $f = 13.56$ MHz. When the impedance at the output port is measured for the three-way combiner using an EM simulator, it is found to be $(29.49 − j0.06)$ Ω at the operational frequency. There is an additional planar circuitry to combine the signals at the output, as shown in Figure 4.52. Each leg of this circuit is connected to the output of the L–C network. The impedance at the interface of each leg after the transformation by the L–C network is $Z_c = 90$ Ω. The phase difference between each leg and the insertion loss introduced by it are adjusted to be minimum. This is required for a properly balanced combiner. Figure 4.56 shows the simulated planar circuitry to combine the output of each L–C network. The simulation results for the planar circuit depicted in Figure 4.56 show that the insertion loss caused by each of the legs is −0.033 dB at 13.56 MHz. The phase difference between each of the legs at the operational frequency is given in Table 4.2.

On the basis of the electromagnetic simulation results, the inductance of the spiral inductor and the shunt capacitance on the L-network are found to be

$$L = 588.6\,\text{nH}, \quad C_4' = 147\,\text{pF}, \quad \text{at } f = 13.56\,\text{MHz} \tag{4.73}$$

The spiral inductor that is simulated in the three-way combiner is shown in Figure 4.57. The simulated results for inductance value versus frequency of spiral

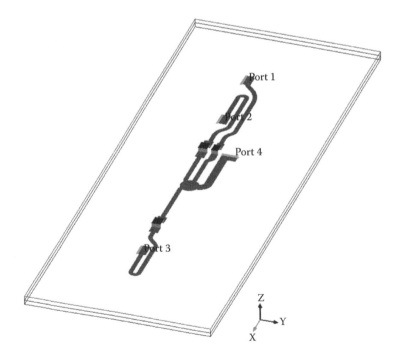

FIGURE 4.56 Simulated planar combining circuitry for three-way combiner.

TABLE 4.2

Phase Information

Ports	Phase Difference (Degree)
1–2	0.13
1–3	−0.11
2–3	−0.24

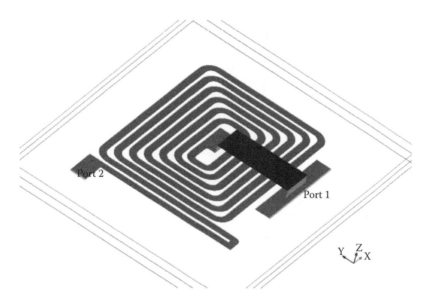

FIGURE 4.57   Spiral inductor that is simulated with method-of-moment-based electromagnetic solver has an inductance of $L = 588.6$ nH.

inductor given in Figure 4.58 show that the resonance occurs at 37.55 MHz. The quality factor and the insertion loss of the spiral inductor are found to be 64.1 and −0.325 dB at the operational frequency, respectively. The simulated quality factor of the spiral inductor versus frequency is given in Figure 4.59. The dimensions of the spiral inductor and the substrate properties are given in Table 4.3. All dimensions are in mil.

The simulated VSWRs at the combined output versus frequency are shown in Figure 4.60. The given impedances on the figures are normalized with respect to the reference impedance. On the basis of the results

Simulated combiner VSWR = 1.695

The picture of the final constructed combiner is shown in Figure 4.61. The measurements are done using an HP-8504A network analyzer. The measured isolation loss and insertion loss for the combiner are illustrated in Figures 4.62 and 4.63 and compared with the simulated results. The isolation loss and insertion loss at the operational frequency are measured to be −17.47 dB and −5.24 dB,

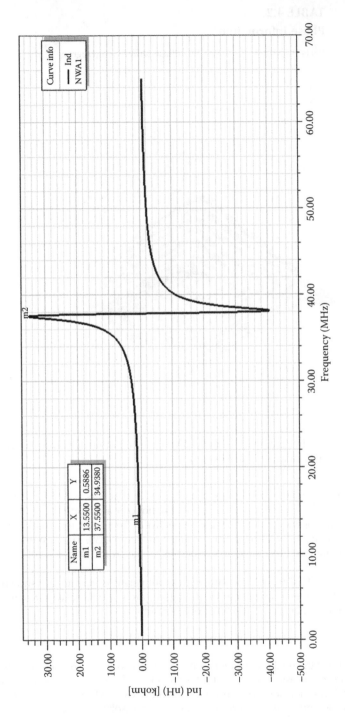

FIGURE 4.58 Simulated spiral inductor inductance versus frequency.

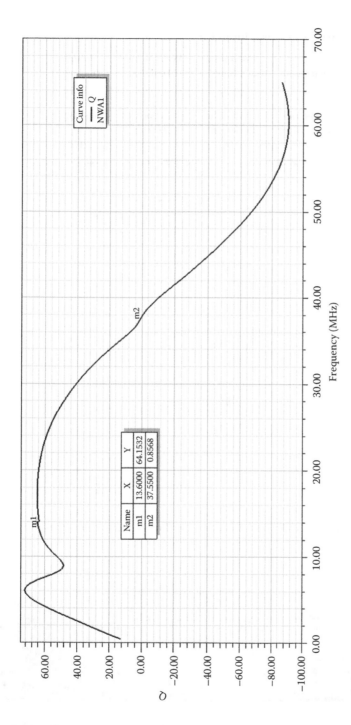

FIGURE 4.59　Simulated spiral inductor quality factor versus frequency.

TABLE 4.3
Physical Dimensions of Spiral Inductor

Trace width w	Spacing s	Horizontal Trace Length l_1	Vertical Trace Length l_2	Copper Thickness t
80	30	1870	1450	4.2

Dielectric Material	Dielectric Permittivity ε_r	Dielectric Thickness h	Number of Turns n	Bridge Height h_b	Bridge Width w_b
Al_2O_3	9.8	100	6.375	100	350

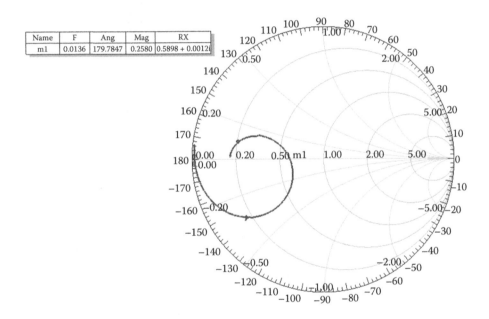

Name	F	Ang	Mag	RX
m1	0.0136	179.7847	0.2580	0.5898 + 0.0012i

FIGURE 4.60 Input VSWR versus frequency for three-way microstrip combiner.

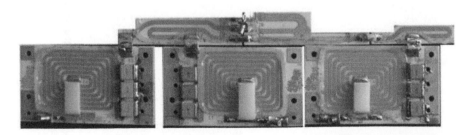

FIGURE 4.61 Three-way combiner implemented in planar form using L-network topology.

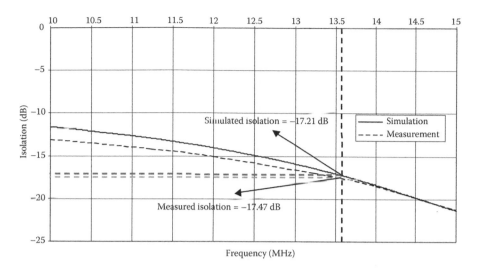

FIGURE 4.62 Measurement results for insertion loss of three-way combiner in planar form using L-network topology.

respectively. The impedance at the output port of the combiner is measured when each distribution port is terminated with 25 Ω impedance. This reflects the real operating condition when each PA module is connected to the combiner. The impedance at the output port under this condition is measured to be (30.23 – j0.06) Ω at the operational frequency. This corresponds to

$$\text{Measured combiner VSWR} = 1.654$$

The measured combiner port impedance is given in Figure 4.64. This value is very close to the targeted impedance value of 30 Ω at the output port.

At the operational frequency, the measured inductance value of the spiral inductor shown in Figure 4.65 is 594.91 nH. The measured inductance value of the spiral inductor versus frequency is given in Figure 4.66. The quality factor and the insertion loss for the spiral inductor are measured to be 58.6 and −0.295 dB, respectively. The measured insertion loss characteristics versus frequency is given in Figure 4.67. As seen in Figures 4.62 and 4.63, the error between the simulation results and the measurement results is small around the operational frequency. This is due to optimization applied at the operational frequency using an electromagnetic simulator. Table 4.4 tabulates the simulated and the measurement results for the three-way combiner.

It is important to note that the analytical model and GUI give results similar with the simulated and measured results when the electrical length is changed from $\theta = 90°$ to $\theta = 65°$ as shown in Figure 4.68. When the adjusted electrical length is inserted in the analytical design, the insertion loss, isolation, and VSWR are in agreement. This deterioration in the response is due to imperfections, loss, and use of L-type network instead of PI-type network in the implementation. However, the use of the L-network simplifies the implementation significantly and reduces the parts counts and associated cost. As a result, the use of the L or PI network topology during transformation depends on the application and requirements.

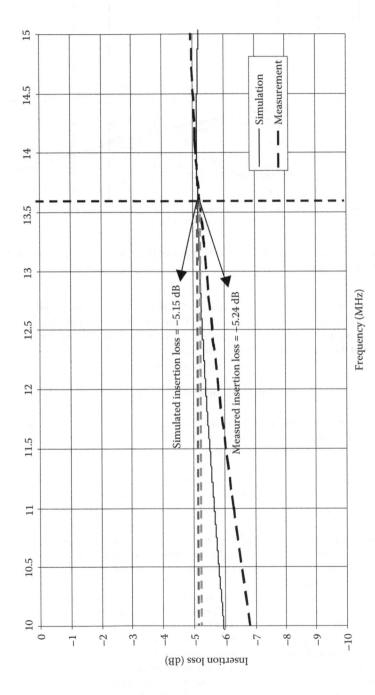

FIGURE 4.63 Measurement results for insertion loss of three-way combiner in planar form using *L*-network topology.

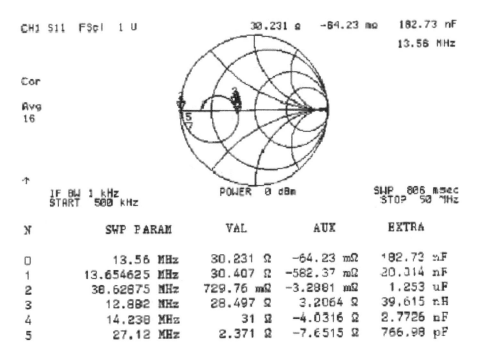

N	SWP PARAM	VAL	AUX	EXTRA
0	13.56 MHz	30.231 Ω	−64.23 mΩ	182.73 nF
1	13.654625 MHz	30.407 Ω	−582.37 mΩ	20.314 nF
2	38.62875 MHz	729.76 mΩ	−3.2881 mΩ	1.253 uF
3	12.882 MHz	28.497 Ω	3.2064 Ω	39.615 nH
4	14.238 MHz	31 Ω	−4.0316 Ω	2.7726 nF
5	27.12 MHz	2.371 Ω	−7.6515 Ω	766.98 pF

FIGURE 4.64   Measured combiner port impedance versus frequency.

Alumina substrate

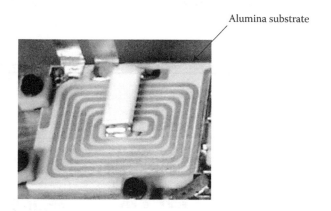

FIGURE 4.65   Spiral inductor on alumina substrate.

4.5   MEASUREMENT SETUP FOR COMBINER/DIVIDER RESPONSE

The important characteristics of the combiners/dividers from a designer point of view are insertion loss, isolation, and VSWRs at the input and output of the system. Isolation and insertion loss are two-port measurements whereas VSWR measurement is a one-port measurement. These design-critical parameters are measured by the network analyzer. Network analyzers have certain reference impedance, which is usually 50 Ω at their measurement ports. Hence, it is important to interface the measurement

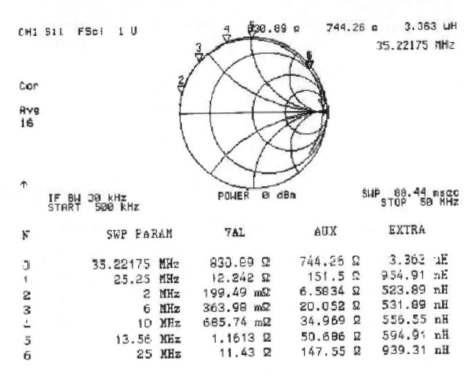

N	SWP PARAM	VAL	AUX	EXTRA
0	35.22175 MHz	830.89 Ω	744.26 Ω	3.363 uH
1	25.25 MHz	12.242 Ω	151.5 Ω	954.91 nH
2	2 MHz	199.49 mΩ	6.5834 Ω	523.89 nH
3	6 MHz	363.98 mΩ	20.052 Ω	531.89 nH
4	10 MHz	685.74 mΩ	34.969 Ω	556.55 nH
5	13.56 MHz	1.1613 Ω	50.686 Ω	594.91 nH
6	25 MHz	11.43 Ω	147.55 Ω	939.31 nH

FIGURE 4.66 Measured inductance value of spiral inductor versus frequency.

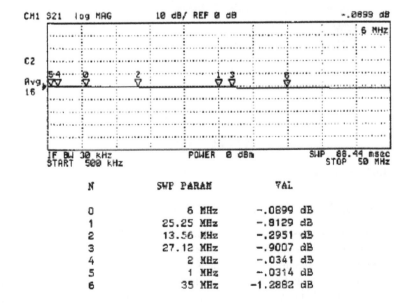

N	SWP PARAM	VAL
0	6 MHz	−.0899 dB
1	25.25 MHz	−.8129 dB
2	13.56 MHz	−.2951 dB
3	27.12 MHz	−.9007 dB
4	2 MHz	−.0341 dB
5	1 MHz	−.0314 dB
6	35 MHz	−1.2882 dB

FIGURE 4.67 Measured insertion loss of spiral inductor versus frequency.

TABLE 4.4

Simulation and Measurement Results

	Measured	Simulated	Measured	Simulated	Measured	Simulated
	Inductance (nH)		Insertion Loss (dB)		Quality Factor	
Spiral inductor	594.91	588.6	−0.295	−0.325	58.6	64
	Capacitance (pF)					
Capacitor	133	147				
	Impedance (Ω)					
Output impedance of the combiner	$30.26 - j0.06$	$29.55 - j0.55$				
	Insertion Loss (dB)		Isolation (dB)			
Combiner	−5.24	−5.25	−17.47	−17.21		

ports of the network analyzers with the same impedance of the device under test (DUT) unless a specific method is used to take into account of reflections at the measurement ports. The concept of the measurement of characteristic responses for power combiners/dividers can be better understood by considering the two-port combiner that is used to combine the output of PA modules as shown in Figure 4.69. In Figure 4.69, there are two PA modules that have two separate amplifiers that are combined through output balun and then fed to the input of two-way combiner.

Isolation of the two-way power combiner shown in Figure 4.69 is experimentally determined by measuring the attenuation between ports A and B while terminating the output port with the required impedance. The transformer is used to interface the network analyzer port impedance to combiner port impedances. This is crucial during measurement since it prevents reflections at measurement ports due to mismatch that affects the accuracy of the measured response. The measurement setup circuit for isolation is shown in Figure 4.70.

The insertion loss is measured between one of the input ports and output ports while terminating all other input ports with required impedances. The measurement circuit for insertion loss is illustrated in Figure 4.71. VSWR measurement is a one-port measurement and it is performed by terminating all other impedances with the required impedance value as shown in Figure 4.72. The measurement methods described in this section are used in the design example given for a three-way power combiner.

4.6 IMPLEMENTATION OF POWER COMBINERS/DIVIDERS USING TRANSFORMER TECHNIQUES

The requirements for high-power and broadband characteristics for combiners/dividers may necessitate the use of conventional and transmission line transformers. On the basis of the windings connection arrangement of the transformers, desired phasing, combining, and dividing the signal with isolation are achieved. The conventional and transmission line transformer design techniques are discussed in detail in Chapter 3 and will be applied in the design of combiners/dividers in this section.

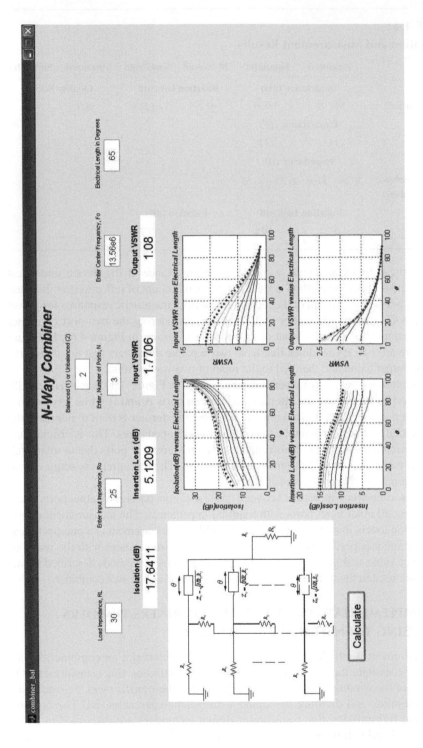

FIGURE 4.68 MATLAB GUI output for three-way unbalanced combiner when $\theta = 65°$.

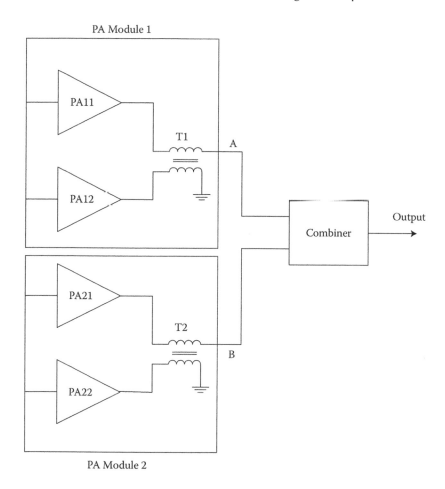

FIGURE 4.69 Two-way combiner implementation for output balun.

Design Example: Four-Way Lumped-Element Divider for RF Power Amplifiers

Design, simulate, and build a broadband four-way divider to be used as test fixture to measure the performance of four-way high-power planar combiner that is used to combine outputs of power amplifiers. The combiner has 25 Ω input port impedances and operates at 13.56 MHz. Measure the response of the four-way divider and determine the amount of attenuation and mismatch introduced by it.

SOLUTION

The bandwidth of the combiner/divider can be increased by increasing the number of the transformers used in the design. The proposed broadband four-way divider is illustrated in Figure 4.73. The topology shown in the implementation of the divider is commonly used for ISM RF applications due to its broadband characteristics. In Figure 4.73, the capacitors are used only as compensation capacitors.

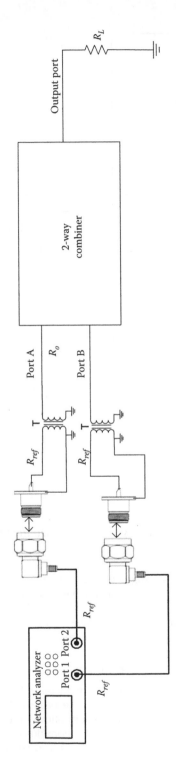

FIGURE 4.70 Isolation measurement setup circuit.

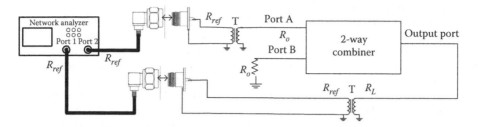

FIGURE 4.71   Insertion loss measurement setup circuit.

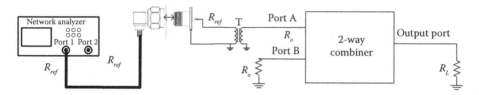

FIGURE 4.72   VSWR measurement setup circuit.

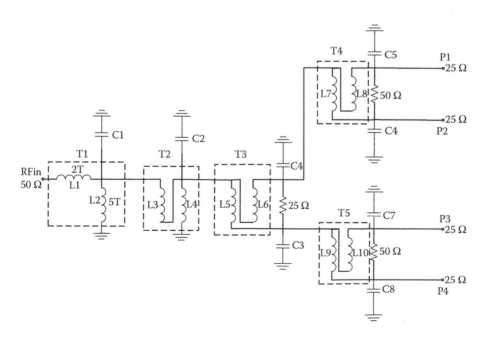

FIGURE 4.73   Four-way divider using transformers.

Use of capacitors is not required when the combiner is being simulated since ideal lumped-element components are used for simulation.

In the divider shown in Figure 4.73, there are two stages that convert the source signal to the desired signal level that will be fed to T3. T1 is the autotransformer that converts 50 Ω source impedance to 25 Ω and T2 is the TLT that transforms 25 Ω impedance to 6.25 Ω output impedance. 6.25 Ω is then transformed into two in-phase signal with 12.5 Ω impedance by another TLT, T3. T4 and T5 are identical TLTs that are used to convert 12.5 Ω impedance to 25 Ω port imped- ance that will be interfaced with the combiner. The balancing resistors between ports are used for TLTs T3–T5 to maintain low VSWR during the operation. The frequency response of the divider is simulated by Ansoft Designer and is shown in Figure 4.74. The theoretical insertion loss for a four-way divider is 6 dB. Insertion loss, isolation, and input VSWR are given in Figures 4.75 through 4.77.

The isolation and insertion loss given in Figures 4.75 and 4.76 illustrate the frequency response of the divider up to 30 MHz. As seen, the insertion loss is maintained across the frequency band that is simulated. The isolation is 40.75 dB at the center frequency and also always better than 26 dB with frequencies up to 30 MHz. The input VSWR of the divider versus frequency within 30 MHz range is given in Figure 4.77.

Since the frequency response of the combiner meets the desired perfor- mance requirement, we can then go ahead and build the power divider circuit. Transformer details are given in Table 4.5.

The constructed four-way divider is shown in Figure 4.78. The response of the four-way divider to determine the attenuation or mismatch introduced by it is measured by connecting two similar divider back to back as shown in Figures 4.79 and 4.80, respectively.

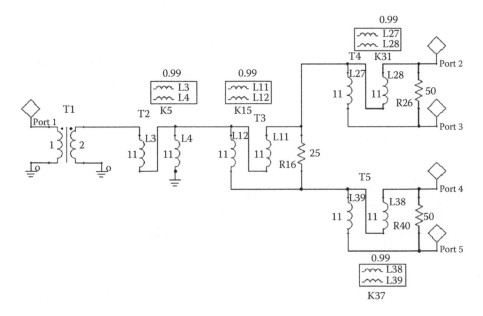

FIGURE 4.74   Simulated four-way divider circuit for frequency response.

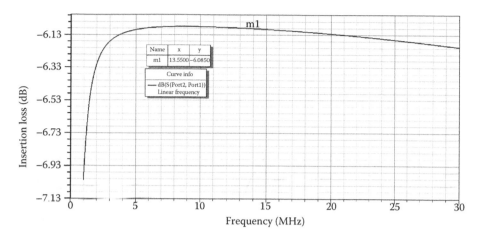

FIGURE 4.75   Simulated insertion loss of four-way divider circuit.

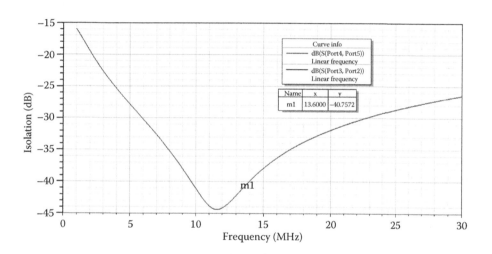

FIGURE 4.76   Simulated isolation of four-way divider circuit.

The four-way divider shown in Figure 4.78 is connected as shown in Figures 4.79 and 4.80 using the second similar set up and measured using the network analyzer for attenuation and mismatch. The insertion loss is measured when the two setup connected as shown in Figure 4.79 is 0.2812 dB.

When the four-way divider that is designed and built in this example is used for the measurement of a four-way combiner for insertion loss, the actual insertion loss of the combiner is found by subtracting the half of divider insertion loss measured and shown in Figure 4.81 from the measured insertion loss value of the combiner. The measured input and output impedances of four way divider versus frequency is given in Figure 4.82.

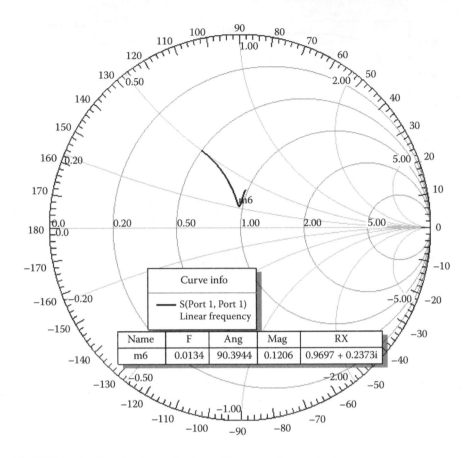

FIGURE 4.77   Simulated insertion loss of four-way divider circuit.

Design Example: Four-Way Microstrip Divider System for Phase Antennas

Design, simulate, and build a microstrip antenna feeder system for four antennas that will be fed from the 50 Ω source at 1 GHz. Each antenna in the array has 100 Ω input impedances. The signal fed into the antennas should have the same amplitude and phase. As a substrate material, use Duroid RT5880 with dielectric constant, $\varepsilon_r = 2.2$, and substrate thickness, $d = 1/16''$. Consider all the effects in the feeder system, including discontinuities due to transition from one microstrip element to another, bends, and so on.

SOLUTION

The design can be realized by using the topology proposed in Figure 4.83. The topology is given in a way that it can be simulated by Ansoft Designer for performance analysis before construction.

In the proposed design, signal from the 50 Ω source is fed to the transmission line with the same characteristic impedance. The signal is then split and sent over the 100 Ω microstrip line and matched to the 50 Ω transmission line via the

TABLE 4.5
Transformer Details Used in Four-Way Divider

Transformer Designation	Transformer Type	Core Type	Core Manufacturer	Manufacturer Part Number	Core Permeability	Core OD (cm)	Core ID (cm)	Core h (cm)	Wire Size	Winding Type	Number of Turns at Input
T1	Auto transformer	Toroid	Magnetics	M2-1319-1C	40	0.78	0.38	0.3	24AWG		2
T2, T3	TLT	Toroid	Ferronics	11-220-PW	40	0.95	0.48	0.32	26AWG	Quad Filar	7
T4, T5	TLT	Toroid	Ferronics	11-220-PW	40	0.95	0.48	0.32	24AWG	Bi Filar	7

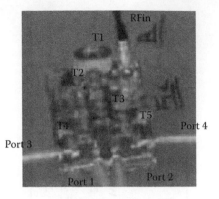

FIGURE 4.78   Constructed four-way divider.

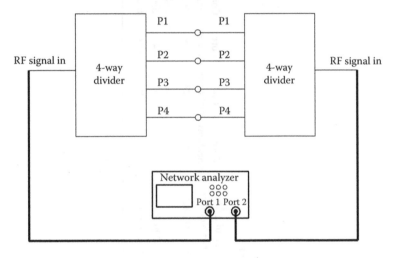

FIGURE 4.79   Measurement setup to determine the attenuation of divider.

quarter wave transformer. Signal on the 50 Ω transmission line is then split equally again and sent to the antenna through the transmission line with the 100 Ω characteristic impedance. The structure is completely symmetrical and theoretically provides equal amplitude and phase on the lines that feed the antennas. The structure can be simulated by Ansoft Designer. However, the accurate response using this simulator is possible only if the discontinuities boxed in Figure 4.83 are taken into account. This can be accomplished by using the components called bend, step, and tee in the circuit simulator section of Ansoft Designer. The distance between antennas is equal to a quarter of a wavelength and is calculated from

$$d = \lambda/4 = 75 \text{ mm}$$

The calculation of the physical dimensions for the corresponding impedances for microstrip transmission lines shown in Figure 4.83 are given by Equations 4.74 through 4.85 as

$$\text{Microstrip } Z_o = 50 \ \Omega, \ W/d > 2$$

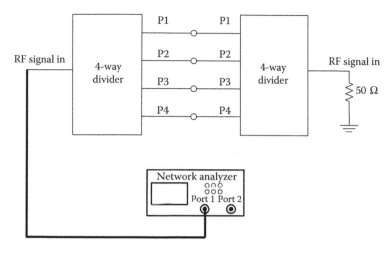

FIGURE 4.80 Measurement setup to determine the mismatch of divider.

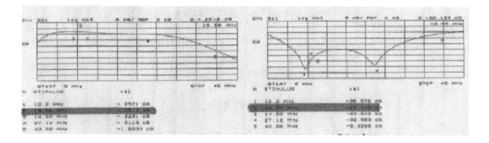

FIGURE 4.81 Measured insertion loss and return loss of four-way divider.

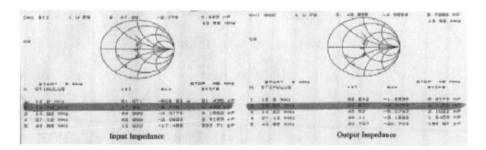

FIGURE 4.82 Measured impedance at the input and output of four-way divider.

$$B = [377\pi/(2 \cdot Z_o \cdot \sqrt{\varepsilon_r})] = 7.98509 \tag{4.74}$$

$$W/d = (2/\pi) \cdot [B - 1 - \ln(2B - 1) + [(\varepsilon_r - 1)/2\varepsilon_r]$$
$$[\ln(B - 1) + 0.39 - (0.61/\varepsilon_r)] = 3.08117 \tag{4.75}$$

$$W = 4.891 \text{ mm} \tag{4.76}$$

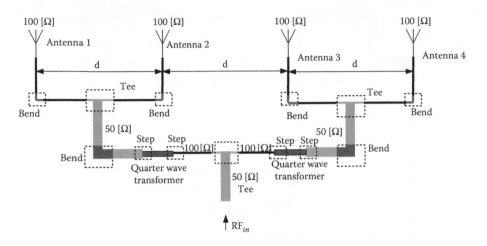

FIGURE 4.83 Proposed four-way microstrip divider for antenna.

Microstrip $-Z_o = 100\ \Omega$, $W/d < 2$

$$A = \{(Z_o/60) \cdot \sqrt{[(\varepsilon_r + 1)/2]}\} + \{[(\varepsilon_r - 1)/(\varepsilon_r + 1)] \cdot [0.23 + (0.11/\varepsilon_r)]\} = 2.21319 \qquad (4.77)$$

$$W/d = (8e^A)/(e^{2A} - 2) = 0.896249 \qquad (4.78)$$

$$W = 1.423\ \text{mm} \qquad (4.79)$$

Microstrip quarter wave transformer $-Z_o$, $W/d < 2$

$$Z_o = \sqrt{(100 \cdot 50)} = 70.71\ \Omega \qquad (4.80)$$

$$A = \{(Z_o/60) \cdot \sqrt{[(\varepsilon_r + 1)/2]}\} + \{[(\varepsilon_r - 1)/(\varepsilon_r + 1)] \cdot [0.23 + (0.11/\varepsilon_r)]\} = 1.59571 \qquad (4.81)$$

$$W/d = (8e^A)/(e^{2A} - 2) = 1.76745 \qquad (4.82)$$

$$W = 2.806\ \text{mm} \qquad (4.83)$$

The transmission line length is found from

$$l = \varphi/[(\sqrt{\varepsilon_e}) \cdot k_o] = 55.67\ \text{mm} \qquad (4.84)$$

where the effective permittivity constant is given by

$$\varepsilon_e = [(\varepsilon_r + 1)/2] + \{[(\varepsilon_r - 1)/2] \cdot \{1/\sqrt{[1 + (12d/W)]}\}\} = 1.81498 \qquad (4.85)$$

k_o is the free space wavenumber and w is the width of the microstrip transmission line. The physical dimensions calculated above are used in the simulation of the divider for the circuit given in Figure 4.84. The initial physical dimensions are optimized and the final physical dimensions are illustrated in the figure.

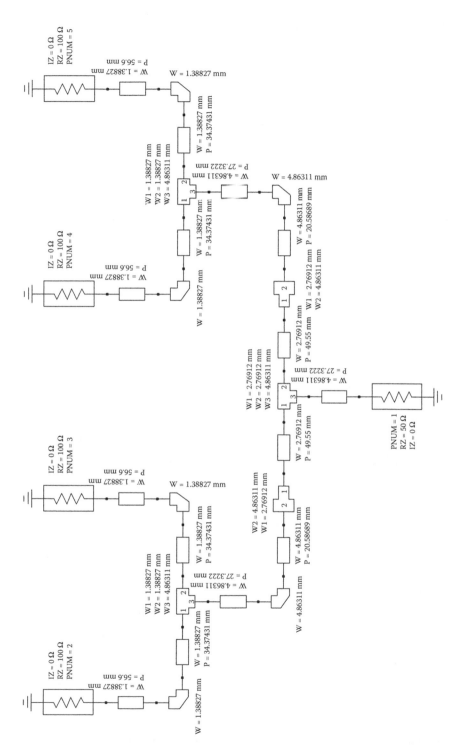

FIGURE 4.84 Simulated optimized four-way microstrip divider for antenna.

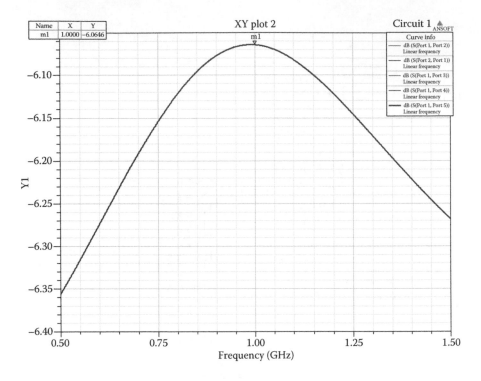

FIGURE 4.85   Simulated insertion loss for microstrip four-way divider.

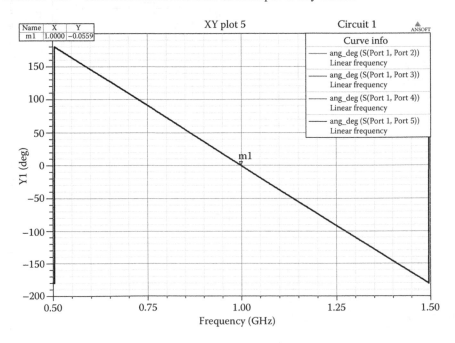

FIGURE 4.86   Simulated phase response for microstrip four-way divider.

The simulated results for insertion loss and return loss are given in Figures 4.85 and 4.86. Insertion loss is found to be 6.06 dB at 1 GHz. This is very close to the response of the expected output of a four-way divider. The phase response in Figure 4.86 illustrates that the phase difference between antenna feeding ports is very small and equal to 0.05° at 1 GHz as desired. The design meets with the specifications outlined and can be implemented. The layout of the divider that will be implemented is shown in Figure 4.87.

The constructed divider for the antenna feeder system is shown in Figure 4.88. When the divider is measured with the network analyzer, the insertion loss and worst-case phase difference between output ports are measured to be 6.12 dB and 7.25°. Slight differences between the simulated and measured results are expected due to imperfections in the measurement setup.

4.7   ANALYSIS AND DESIGN OF PHASE INVERTER USING TLT

When the basic TLT block is implemented as shown in Figure 4.89, it operates as a phase inverter.

A similar circuit analysis that is performed for balun can be done for phase inverter as illustrated in Figure 4.89. In the analysis of the phase inverter circuit shown in Figure 4.89, we use the constraints given in Equations 4.86 through 4.88

$$L_1 = L_2 = L \tag{4.86}$$

$$I_1 = -I_2 = I \tag{4.87}$$

$$V_1 = -V_2 = V \tag{4.88}$$

The application of KVL_1 and KVL_2 gives

$$V_s - I_1 R_s - V = 0 \tag{4.89}$$

$$V_L - V = 0 \tag{4.90}$$

Since

$$V_L = -I_2 R_L, \quad \text{then } I_2 = -\frac{V_L}{R_L} \quad \text{or} \quad I_1 = \frac{V_L}{R_L} \tag{4.91}$$

Substitution of Equation 4.91 into Equation 4.89 gives

$$V_s - \frac{V_L R_s}{R_L} - V_L = 0 \quad \text{or} \quad V_s - V_L\left(\frac{R_s}{R_L} + 1\right) = 0$$

which leads to

$$V_L\left(\frac{R_s + R_L}{R_L}\right) = V_s \quad \text{or} \quad V_L = V_s\left(\frac{R_L}{R_s + R_L}\right) \tag{4.92}$$

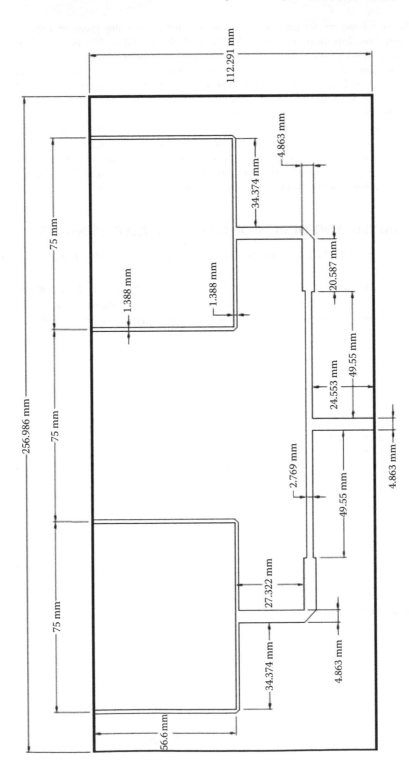

FIGURE 4.87 The layout of four-way microstrip divider.

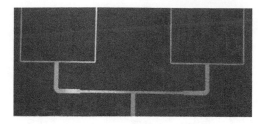

FIGURE 4.88   The constructed four-way microstrip divider for antenna feeder system.

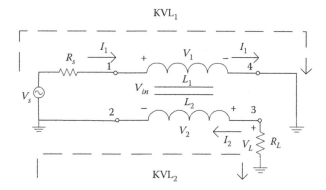

FIGURE 4.89   Implementation of phase inverter in Pspice.

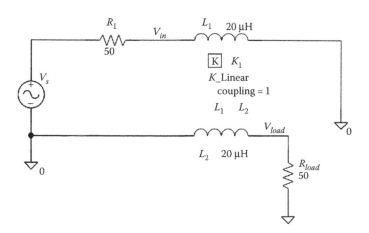

FIGURE 4.90   Implementation of phase inverter in Pspice.

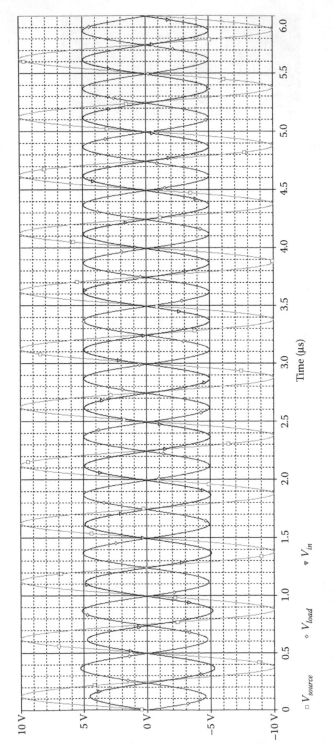

FIGURE 4.91   Pspice simulation results for alternative balun implementation.

As a result, when $R_s = R_L$, the output voltage is

$$V_L = \frac{V_s}{2} \quad \text{or} \quad V_L = V_{in} \tag{4.93}$$

Hence

$$V_{R_L} = -V_L = -V_{in} \tag{4.94}$$

Please note that due to the direction of the current and use of passive sign convention, the voltage across the load resistor is 180° out of phase than the input voltage. The circuit in Figure 4.89 is simulated with Pspice as shown in Figure 4.90 and illustrates and confirms the relations shown by Equations 4.89 through 4.94.

The netlist of the circuit is

```
Kn_K1 L_L1 L_L2 1
L_L1         Vin 0 20uH
L_L2         0 Vload 20uH
R_Rload 0 Vload 50
R_R1         Vsource Vin 50
V_Vs         Vsource 0 +SIN 0 10 2meg 0 0 0
.Tran 0 ns 10us 0 1n
.Probe
.End
```

Pspice simulation results for the phase inverter are shown in Figure 4.91.

REFERENCES

1. E.J. Wilkinson. An n-way hybrid power divider, *IRE Transactions on Microwave Theory and Techniques*, MTT-8, 116–118, 1960.
2. A.D. Saleh. Planar electrically symmetric n-way hybrid power dividers and combiners, *IEEE Transactions on Microwave Theory and Techniques*, MTT-28, 555–563, 1980.
3. H. Howe, Jr. Simplified design of high power, N-way, in-phase power divider/combiners, *Microwave Journal*, December, 51–53, 1979.
4. Tang, X. and K. Mouthaan, Analysis and design of compact two-way Wilkinson power dividers using coupled lines, *APMC*, Singapore, pp. 1319–1322, 2009.
5. R. Knochel and B. Mayer, Broadband printed circuit 0°/180° couplers and high power in phase power dividers, *1990 IEEE MTT-S International Microwave Symposium Digest*, 1, 471–474,1990.
6. K.J. Russel. Microwave power combining techniques, *IEEE Transactions on Microwave Theory and Techniques*, MTT-27, 472–478, 1979.
7. V.F. Fusco and S.B.D. O'Caireallain. Lumped element hybrid networks for GaAs MMICs, *Microwave and Optical Technology Letters*, (2)(1), 19–23, 1989.
8. D.M. Kinman, D.J. White, and M. Afendykiw. Symmetrical combiner analysis using S-parameters, *IEEE Transactions on Microwave Theory and Techniques*, MTT-30, 268–277, 1982.

9. R.L. Ernst, R.L. Camisa, and A. Presser, Graceful degradation properties of matched n-port power amplifier combiners, *International Microwave Symposium Digest,* 174:177, 1977.

10. U.H. Gysel. A new-way power divider/combiner suitable for high power applications, *IEEE MTT-S International Microwave Symposium Digest,* 75, 116–118, 1975.

11. B. Schüppert. Microstrip/slotline transitions: Modeling and experimental investigation, *IEEE Transactions on Microwave Theory and Techniques,* 36(8), 1272–1282, 1988.

12. M.M. Elsbury 1, P.D. Dresselhaus, S.P. Benz, and Z. Popovic, Integrated broadband lumped-element symmetrical-hybrid N-way power dividers, *International Microwave Symposium Digest,* 997–1000, 2009.

5 MF-UHF Directional Coupler Design Techniques

5.1 INTRODUCTION

Directional couplers are used for many RF and microwave applications to function as attenuator, power splitter, hybrid junction, and most commonly a sampling device for measuring forward and backward waves on a TL [1–11]. Directional couplers can be implemented using TLs such as microstrip and stripline or lumped elements depending on the application and operational frequency. TL directional couplers give lower profile and are very popular due to their several advantages, including manufacturability, repeatability, and low cost. In the design of directional couplers when they are used as sampling devices, the key design parameters that the designer needs to give special attention are directivity, VSWR, coupling level, and insertion loss. Directional coupler is a passive four-port device that is shown in Figure 5.1.

The scattering matrix for a four-port device is given by

$$S = \begin{bmatrix} S_{11} & S_{12} & S_{13} & S_{14} \\ S_{21} & S_{22} & S_{23} & S_{24} \\ S_{31} & S_{31} & S_{33} & S_{34} \\ S_{41} & S_{42} & S_{43} & S_{44} \end{bmatrix} \tag{5.1}$$

If the device is matched at all ports, then

$$S_{11} = S_{22} = S_{33} = S_{44} = 0 \tag{5.2}$$

Since the directional coupler is assumed to be lossless, then the scattering matrix has to satisfy

$$S^{\dagger} S = I \tag{5.3}$$

$\dagger$ denotes the conjugate transpose of the matrix and I is the unit matrix. Executing Equation 5.3 gives

$$S_{14}^{*} \left(|S_{13}|^2 - |S_{24}|^2 \right) = 0 \tag{5.4}$$

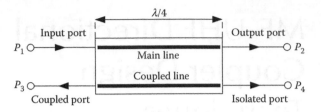

FIGURE 5.1 Conventional directional coupler.

$$S_{23}\left(\left|S_{12}\right|^2 - \left|S_{34}\right|^2\right) = 0 \tag{5.5}$$

which leads to $S_{14} = S_{23} = 0$. When we assume the network symmetrical for the device such as directional coupler, then

$$S_{14} = S_{41} = S_{23} = S_{32} = 0 \tag{5.6}$$

Then, the resulting equations from Equation 5.3 can be written as

$$\left|S_{12}\right|^2 + \left|S_{13}\right|^2 = 1 \tag{5.7}$$

$$\left|S_{12}\right|^2 + \left|S_{24}\right|^2 = 1 \tag{5.8}$$

$$\left|S_{13}\right|^2 + \left|S_{34}\right|^2 = 1 \tag{5.9}$$

$$\left|S_{24}\right|^2 + \left|S_{34}\right|^2 = 1 \tag{5.10}$$

Equations 5.7 through 5.10 lead to

$$\left|S_{13}\right| = \left|S_{24}\right|, \quad \left|S_{12}\right| = \left|S_{34}\right| \tag{5.11}$$

The scattering matrix for the symmetrical, lossless coupler can now be written as

$$S = \begin{bmatrix} 0 & \alpha & j\beta & 0 \\ \alpha & 0 & 0 & j\beta \\ j\beta & 0 & 0 & \alpha \\ 0 & j\beta & \alpha & 0 \end{bmatrix} \tag{5.12}$$

where

$$S_{12} = S_{34} = \alpha, \quad S_{13} = S_{24} = j\beta \tag{5.13}$$

The coupling level for directional couplers is the main parameter in the device's performance. It varies with the frequency. The requirement for directional coupler is to minimize the variation of the coupling level within the frequencies of operation. The coupling of the directional coupler shown in Figure 5.1 can be defined as

$$\text{Coupling level (dB)} = 10\log\left(\frac{P_1}{P_3}\right) = -20\log(\beta) \tag{5.14}$$

where P_1 is the input power and P_3 is the output power at the coupled port. Isolation of a directional coupler can be defined as the difference in signal levels between the input port and the isolated port when the two output ports are matched loads. The requirement is to have high isolation between the input and isolated ports so that the signal purity can be maintained. Ideal couplers have infinite isolation. The isolation level can be then defined as

$$\text{Isolation level (dB)} = 10\log\left(\frac{P_1}{P_4}\right) = -20\log\left(|S_{14}|\right) \tag{5.15}$$

Directivity is related to isolation and can be calculated using

$$\text{Directivity level (dB)} = 10\log\left(\frac{P_3}{P_4}\right) = 20\log\left(\frac{\beta}{|S_{14}|}\right) \tag{5.16}$$

The directivity is a measure of how well the coupler can isolate two signals. Higher directivity results in better measurement accuracy. As a result, it is the most critical design parameter for accurate measurement systems.

In this chapter, several microstrip and lumped element directional coupler topologies will be analyzed and design examples will be given.

5.2 MICROSTRIP DIRECTIONAL COUPLERS

Consider the geometry of a symmetrical microstrip directional coupler as shown in Figure 5.2.

In practice, port termination impedances, coupling level, and operational frequency are the input design parameters that are being used to realize couplers. The matched system is accomplished when the characteristic impedance

$$Z_o = \sqrt{Z_{oe}Z_{oo}} \tag{5.17}$$

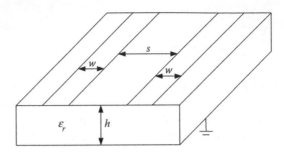

FIGURE 5.2 Geometry of symmetrical two-line microstrip directional coupler.

is equal to the port impedance. In Equation 5.2, Z_{oe} and Z_{oo} are even and odd mode impedances, respectively. The even and odd impedances, Z_{oe} and Z_{oo}, of the microstrip coupler given in Figure 5.2 can be found from

$$Z_{oe} = Z_o \sqrt{\frac{1 + 10^{C/20}}{1 - 10^{C/20}}} \tag{5.18}$$

$$Z_{oo} = Z_o \sqrt{\frac{1 - 10^{C/20}}{1 + 10^{C/20}}} \tag{5.19}$$

where C is the forward coupling requirement and is given in decibels. The physical dimensions of the directional coupler are found using the synthesis method. Application of the synthesis method gives the spacing ratio s/h of the coupler in Figure 5.1 as

$$s/h = \frac{2}{\pi} \cosh^{-1} \left[\frac{\cosh\left[\pi/2(w/h)_{se}\right] + \cosh\left[\pi/2(w/h)'_{so}\right] - 2}{\cosh\left[\pi/2(w/h)'_{so}\right] - \cosh\left[\pi/2(w/h)_{se}\right]} \right] \tag{5.20}$$

$(w/h)_{se}$ and $(w/h)_{so}$ are the shape ratios for the equivalent single case corresponding to even and odd mode geometries, respectively. $(w/h)'_{so}$ is the second term for the shape ratio. (w/h) is the shape ratio for the single microstrip line and it is expressed as

$$\frac{w}{h} = \frac{8\sqrt{\left[\exp\left((R/42.4)\sqrt{\varepsilon_r + 1}\right) - 1\right]\left(7 + (4/\varepsilon_r)/11\right) + \left(1 + (1/\varepsilon_r)/0.81\right)}}{\exp\left((R/42.4)\sqrt{\varepsilon_r + 1}\right) - 1} \tag{5.21}$$

where

$$R = \frac{Z_{oe}}{2} \quad \text{or} \quad R = \frac{Z_{oo}}{2} \tag{5.22}$$

Z_{ose} and Z_{oso} are the characteristic impedances corresponding to single microstrip shape ratios $(w/h)_{se}$ and $(w/h)_{so}$, respectively. They are given as

$$Z_{ose} = \frac{Z_{oe}}{2} \tag{5.23}$$

$$Z_{oso} = \frac{Z_{oo}}{2} \tag{5.24}$$

and

$$(w/h)_{se} = (w/h)\Big|_{R = Z_{ose}} \tag{5.25}$$

$$(w/h)_{so} = (w/h)\Big|_{R = Z_{oso}} \tag{5.26}$$

The term $(w/h)'_{so}$ in Equation 5.20 is given as

$$\left(\frac{w}{h}\right)'_{so} = 0.78\left(\frac{w}{h}\right)_{so} + 0.1\left(\frac{w}{h}\right)_{se} \tag{5.27}$$

After the spacing ratio s/h for the coupled lines is found, we can proceed to find w/h for the coupled lines. The shape ratio for the coupled lines is

$$\left(\frac{w}{h}\right) = \frac{1}{\pi}\cosh^{-1}(d) - \frac{1}{2}\left(\frac{s}{h}\right) \tag{5.28}$$

where

$$d = \frac{\cosh\left[(\pi/2)(w/h)_{se}\right](g + 1) + g - 1}{2} \tag{5.29}$$

$$g = \cosh\left[\frac{\pi}{2}\left(\frac{s}{h}\right)\right] \tag{5.30}$$

The physical length of the directional coupler is obtained using

$$l = \frac{\lambda}{4} = \frac{c}{4f\sqrt{\varepsilon_{eff}}} \tag{5.31}$$

where $c = 3 * 10^8$ m/s and f is the operational frequency in hertz. Hence, the length of the directional coupler can be found if the effective permittivity constant ε_{eff} of the coupled structure shown in Figure 5.1 is known. ε_{eff} can be found using the relation given in Equation 5.16 as

$$\varepsilon_{eff} = \left[\frac{\sqrt{\varepsilon_{effe}} + \sqrt{\varepsilon_{effo}}}{2} \right]^2 \qquad (5.32)$$

ε_{effe} and ε_{effo} are the effective permittivity constants of the coupled structure for odd and even modes, respectively. ε_{effe} and ε_{effo} depend on the even and odd mode capacitances C_e and C_o as

$$\varepsilon_{effe} = \frac{C_e}{C_{e1}} \qquad (5.33)$$

$$\varepsilon_{effo} = \frac{C_o}{C_{o1}} \qquad (5.34)$$

$C_{e1,o1}$ is the capacitance with air as dielectric. All the capacitances are given as capacitance per unit length. The even mode capacitance C_e is

$$C_e = C_p + C_f + C_f' \qquad (5.35)$$

The capacitances in the even mode for the coupled lines can be visualized as shown in Figure 5.3a.

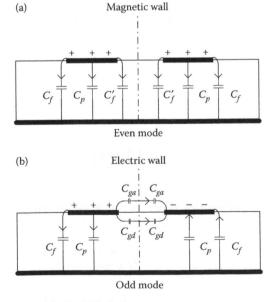

FIGURE 5.3   Coupled lines mode representation. (a) Even mode and (b) odd mode.

C_p is the parallel plate capacitance and is defined as

$$C_p = \varepsilon_0 \varepsilon_r \frac{w}{h} \tag{5.36}$$

where w/h is found in the previous section. C_f is the fringing capacitance due to the microstrip being taken alone as if it were a single strip. That is equal to

$$C_f = \frac{\sqrt{\varepsilon_{seff}}}{2cZ_o} - \frac{C_p}{2} \tag{5.37}$$

Here, ε_{seff} is the effective permittivity constant of a single-strip microstrip. It can be expressed as

$$\varepsilon_{seff} = \frac{\varepsilon_r + 1}{2} - \frac{\varepsilon_r - 1}{2} F(w/h) \tag{5.38}$$

where

$$F(w/h) = \begin{cases} (1 + 12h/w)^{-1/2} + 0.041(1 - w/h)^2 & \text{for } \left(\dfrac{w}{h} \le 1\right) \\ (1 + 12h/w)^{-1/2} & \text{for } \left(\dfrac{w}{h} \ge 1\right) \end{cases} \tag{5.39}$$

C_f' is given by the following equation:

$$C_f' = \frac{C_f}{1 + A(h/s)\tanh(10s/h)} \left(\frac{\varepsilon_r}{\varepsilon_{seff}}\right)^{1/4} \tag{5.40}$$

and

$$A = \exp\left[-0.1\exp\left(2.33 - 1.5\frac{w}{h}\right)\right] \tag{5.41}$$

The odd mode capacitance C_o is

$$C_o = C_p + C_f + C_{ga} + C_{gd} \tag{5.42}$$

The capacitances in the odd mode for the coupled lines can be visualized as shown in Figure 5.3b. C_{ga} is the capacitance term in the odd mode for the fringing field across the gap in the air region. It can be written as

$$C_{ga} = \varepsilon_0 \frac{K(k')}{K(k)} \tag{5.43}$$

where

$$\frac{K(k')}{K(k)} = \begin{cases} \dfrac{1}{\pi} \ln\left[2\dfrac{1+\sqrt{k'}}{1-\sqrt{k'}}\right], & 0 \le k^2 \le 0.5 \\ \dfrac{\pi}{\ln\left[2\dfrac{1+\sqrt{k'}}{1-\sqrt{k'}}\right]}, & 0.5 \le k^2 \le 1 \end{cases} \tag{5.44}$$

and

$$k = \frac{(s/h)}{(s/h) + (2w/h)} \tag{5.45}$$

$$k' = \sqrt{1 - k^2} \tag{5.46}$$

C_{gd} represents the capacitance in the odd mode for the fringing field across the gap in the dielectric region. It can be found using

$$C_{gd} = \frac{\varepsilon_0 \varepsilon_r}{\pi} \ln\left\{\coth\left(\frac{\pi}{4}\frac{s}{h}\right)\right\} + 0.65 C_f \left[\frac{0.02}{(s/h)}\sqrt{\varepsilon_r} + \left(1 - \frac{1}{\varepsilon_r^2}\right)\right] \tag{5.47}$$

Since

$$Z_{oe} = \frac{1}{c\sqrt{C_e C_{e1}}} \tag{5.48}$$

$$Z_{oo} = \frac{1}{c\sqrt{C_o C_{o1}}} \tag{5.49}$$

Then, we can write

$$C_{e1} = \frac{1}{c^2 C_e Z_{oe}^2} \tag{5.50}$$

$$C_{o1} = \frac{1}{c^2 C_o Z_{oo}^2} \tag{5.51}$$

Substituting Equations 5.35, 5.32, 5.40, and 5.51 into Equations 5.34 and 5.35 gives the even and odd mode effective permittivities ε_{effo}. When Equations 5.34 and 5.35 are substituted into Equation 5.33, we can find the effective permittivity constant ε_{eff} of the coupled structure. Now, Equation 5.31 can be used to calculate the physical length of the directional coupler at the operational frequency. On the basis of the formulation and design procedure outlined above, the design curves for microstrip symmetrical coupler for several popular RF materials such as alumina, Teflon, RO4003, FR4, and RF-60 are calculated and given in by the design charts shown in Figures 5.4 through 5.6.

In addition, the physical length of the coupler lines are also calculated and the design curves are obtained for the commonly used RF materials for several coupling levels versus frequency and are given by Figures 5.7 through 5.9.

The analytical values obtained and given in the design curves are used as physical dimensions for the symmetrical coupler that are simulated by the electromagnetic

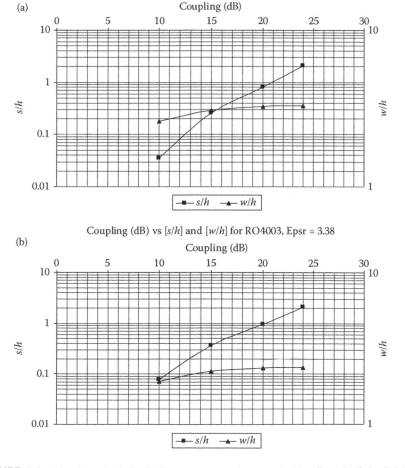

FIGURE 5.4 (a) w/h and s/h for Teflon versus coupling level. (b) w/h and s/h for RO4003 versus coupling level.

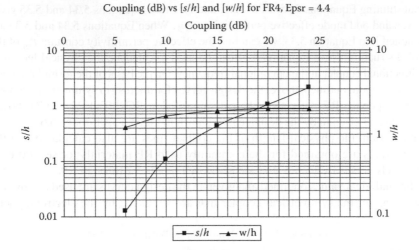

FIGURE 5.5   *w/h* and *s/h* for FR4 versus coupling level.

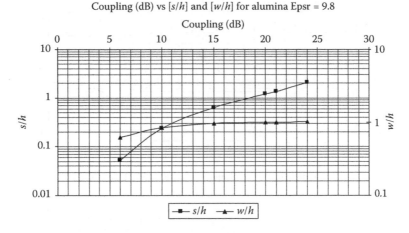

FIGURE 5.6   *w/h* and *s/h* for alumina versus coupling level.

simulator, Ansoft Designer. The error between the analytical and simulated values are shown in Figure 5.10. The simulation is performed at 300 MHz and the error is calculated using the equation in

$$\text{Error } (\%) = \frac{C_{cal}(\text{dB}) - C_{sim}(\text{dB})}{C_{cal}(\text{dB})} \times 100 \qquad (5.52)$$

C_{cal} (dB) and C_{sim} (dB) are the calculated and the simulated coupling levels, respectively. The worst error is equal to 3.8% and occurs when the lower-permittivity material Teflon is used as substrate material. The smallest error is observed with the

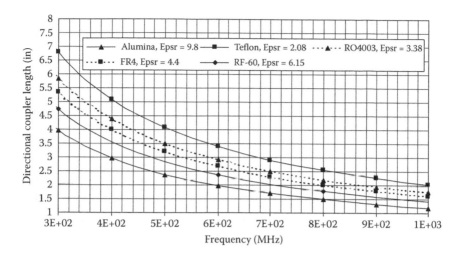

FIGURE 5.7 Physical length of the coupler line versus frequency for 10 dB coupling level.

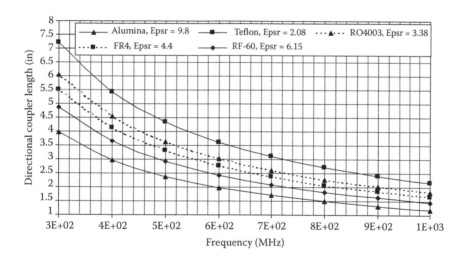

FIGURE 5.8 Physical length of the coupler line versus frequency for 15 dB coupling level.

high-permittivity material alumina—$\varepsilon_r = 9.8$ at −20 dB coupling and equal to 0.2%. As a result, the proposed method to design symmetrical couplers gives very close results to the simulated results.

At this point, MATLAB® GUI has been developed for designers to develop, design, and implement symmetrical couplers in conjunction with simulators. The screen shot of MATLAB GUI is illustrated in Figure 5.11. In the program, coupling level, port impedance, substrate permittivity constant, and operational frequency are used as input. These parameters are the key design parameters that are used as input in the program to design the coupler. The program then gives

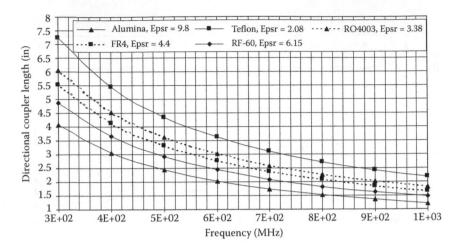

FIGURE 5.9   Physical length of the coupler line versus frequency for 20 dB coupling level.

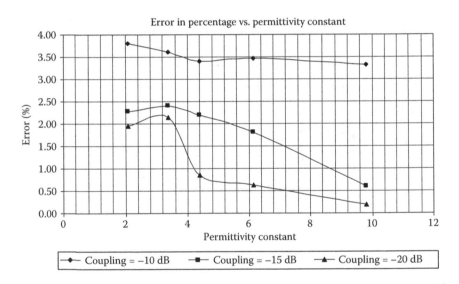

FIGURE 5.10   Error between the simulated and analytical results for coupling level at 300 MHz.

the specific physical dimensions at the frequency of interest and also provides design curves for several parameters, including shape ratio, spacing ratio, odd and even mode capacitances, coupler length, and effective permittivity of the structure versus substrate permittivity for the given coupling level and operational frequency.

Overall, the design of a symmetrical microstrip directional coupler begins with the use of analytical formulation outlined or the use of MATLAB GUI to obtain the physical parameters. The second stage in the design procedure is using

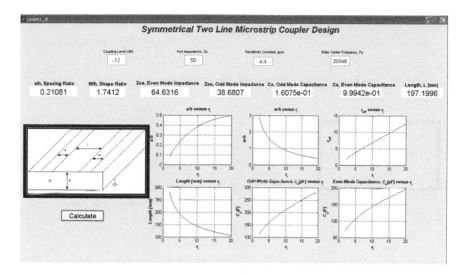

FIGURE 5.11 MATLAB GUI screen shot for symmetrical two-line microstrip coupler design.

the physical values obtained as base values for electromagnetic simulation of the structure. This design procedure shortens the design time and increases the accuracy tremendously.

Design Example: Symmetrical Microstrip Directional Coupler

Design, simulate, and build a 15 dB microstrip symmetrical coupler using Teflon—$\varepsilon_r = 2.08$ and FR4—$\varepsilon_r = 4.4$ at 300 MHz. The thicknesses of substrates are 90 and 120 mil for Teflon and FR4, respectively. Compare, measure, simulate, and analyze the results.

SOLUTION

MATLAB GUI is used to obtain the physical dimensions of the coupler as described earlier using the formulation outlined. The values obtained for the specified coupler using Teflon and FR4 at 300 MHz for 15 dB coupling are illustrated in Figures 5.12 and 5.13.

Now, the couplers are simulated using the physical values given in Figures 5.12 and 5.13. The simulated structure for Teflon is shown in Figure 5.14.

The simulated coupling, isolation, and directivity versus frequency are given by Figures 5.15 through 5.17.

The simulated coupling level is obtained to be 14.659 dB at 300 MHz. The simulated isolation and directivity levels for the coupler are 27.93 and 13.27 dB, respectively. Since the simulated coupling level is in agreement with the calculated value, the structure has been built and the coupling level is measured using HP8751A network. The coupler's frequency response is obtained between 1 and 400 MHz as shown in Figure 5.18.

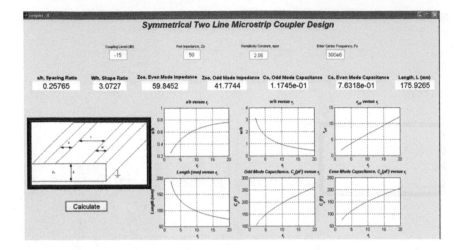

FIGURE 5.12 Analytical values of symmetrical two-line microstrip coupler for Teflon.

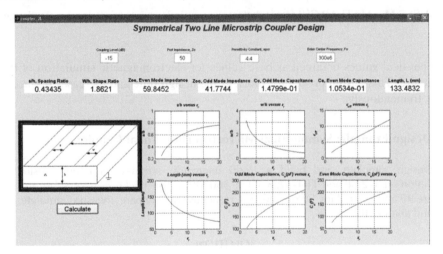

FIGURE 5.13 Analytical values of symmetrical two-line microstrip coupler for FR4.

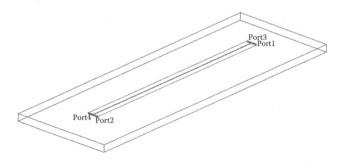

FIGURE 5.14 Simulated two-line symmetrical coupler using Teflon at 300 MHz for 10 dB coupling.

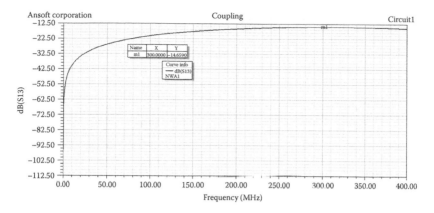

FIGURE 5.15 Simulated coupling level for two-line symmetrical coupler using Teflon at 300 MHz.

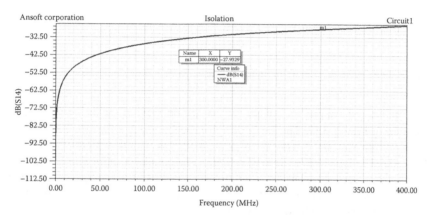

FIGURE 5.16 Simulated isolation level for two-line symmetrical coupler using Teflon at 300 MHz.

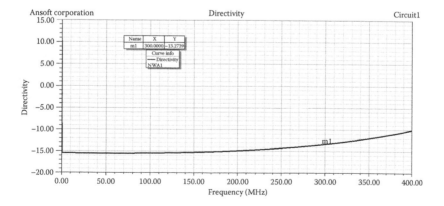

FIGURE 5.17 Simulated directivity level for two-line symmetrical coupler using Teflon at 300 MHz.

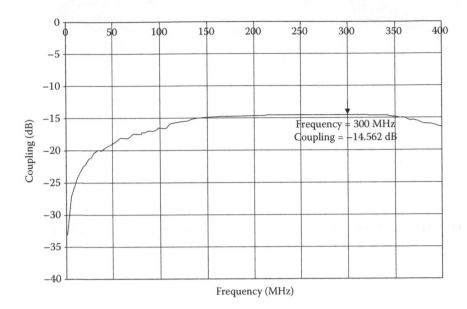

FIGURE 5.18 Measured coupling level for two-line symmetrical coupler using Teflon at 300 MHz.

The measured coupling level for Teflon at 300 MHz is equal to 14.562 dB as illustrated in Figure 5.17. The same design procedure is applied for FR4 and simulation and measurement have been performed. The simulated coupling level for FR4 is found to be −14.671 dB at 300 MHz. The coupler is built and its frequency response for the coupling level is given in Figure 5.19. The coupling level is found to be 14.679 dB at 300 MHz. Overall, the worst error between the measured results and the analytical results is found to be within 3%.

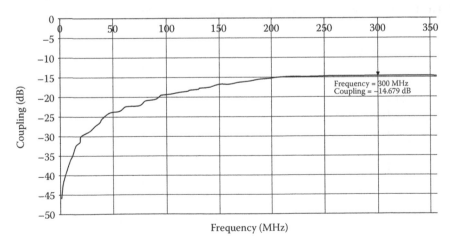

FIGURE 5.19 Measured coupling level for two-line symmetrical coupler using FR4 at 300 MHz.

5.3 MULTILAYER DIRECTIONAL COUPLERS

High directivity in the design of directional couplers can be obtained using multilayer configuration. The asymmetric coupled line is matched and perfectly isolated if

$$k_L = k_C \tag{5.53}$$

and

$$Z_i = \sqrt{\frac{L_i}{C_i}}, \quad i = 1,2 \tag{5.54}$$

where

$$k_L = \frac{L_m}{\sqrt{L_1 L_2}} \tag{5.55}$$

and

$$k_C = \frac{C_m}{\sqrt{C_1 C_2}} \tag{5.56}$$

k_L is defined as the inductive coupling coefficient, k_C is defined as the capacitive coupling coefficient, Z_i is the characteristic impedance of the terminating line, L_1, L_2 are the self-inductances, L_m is the mutual inductance, C_1, C_2 are the self-capacitances, and C_m is the mutual capacitance of the coupled lines illustrated in Figure 5.20. The physical length of the coupler is found from Equation 5.31.

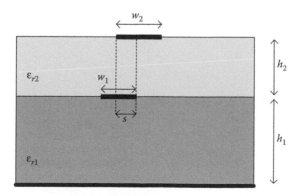

FIGURE 5.20   The geometry of a directional coupler using multilayer asymmetric coupled lines.

Example: Directivity Improvement with Multilayer Configuration

Design a 10 dB symmetrical microstrip two-line directional coupler at 300 MHz using analytical method and simulate it with the values obtained for coupling and directivity levels. Use Teflon material with $\varepsilon_r = 2.08$ with 120 mil thickness as a substrate. Then, modify your design to obtain two-layer configuration and show directivity improvement over frequency.

SOLUTION

The design has been created using MATLAB GUI and design output parameters are shown in Figure 5.21. These parameters will be used in the simulation of the coupler. Spacing ratio, s/h, and shape ratio, w/h, at 300 MHz are found to be 0.035 and 2.61, respectively. The design curves for odd and even mode capacitance are given in Figure 5.21. The physical values obtained are used in electromagnetic Ansoft Designer, and simulation has been performed. The simulated coupler is shown in Figure 5.22. The simulated coupling and directivity levels are

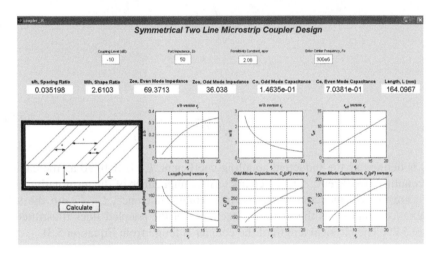

FIGURE 5.21   10 dB microstrip symmetrical directional coupler.

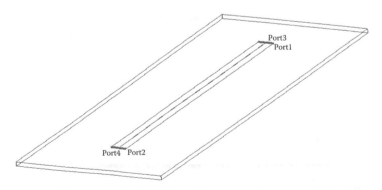

FIGURE 5.22   Simulated 10 dB microstrip symmetrical directional coupler.

found to be 9.75 and 22.26 dB as shown in Figures 5.23 and 5.24. 10 dB coupler design simulation shows that although coupling-level frequency versus frequency is maintained within 600 MHz range, the directivity levels becomes worse as the frequency increases. This is a direct result of poor isolation for higher frequencies.

The structure is now modified as two-layer configuration. The first and second layers are taken to be 60-mil-thick Teflon. The main line is placed on the top layer, which is 120 mil away from the ground plane whereas the coupled line is placed in the second layer, which is 60 mil away from the ground plane. The physical dimensions of the coupled lines are kept the same. The two-layer coupled structure that is simulated is given in Figure 5.25.

The coupling level versus frequency for the two-layer structure is shown in Figure 5.26. As seen, the coupling level with this symmetric structure is 12.3375 dB. The coupling can be brought back to 10 dB level by increasing the thickness of the first layer by 90 mil and keeping everything else constant in the coupler as illustrated in Figure 5.27. The directivity level versus frequency for the modified two-layer coupler

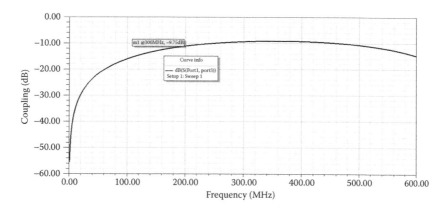

FIGURE 5.23   Simulated coupling level for 10 dB microstrip symmetrical directional coupler.

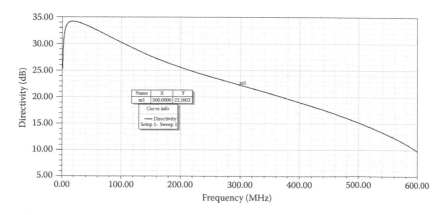

FIGURE 5.24   Simulated directivity level for 10 dB microstrip symmetrical directional coupler.

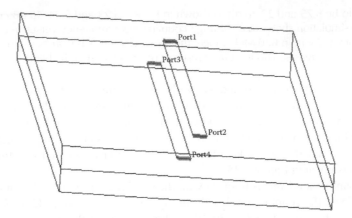

FIGURE 5.25   Simulated multilayer 10 dB microstrip symmetrical directional coupler.

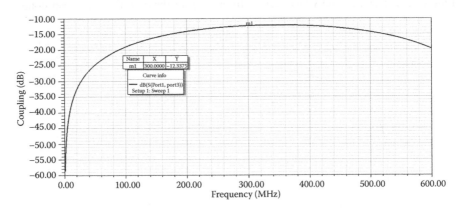

FIGURE 5.26   Simulated coupling level of two-layer 10 dB microstrip symmetrical directional coupler.

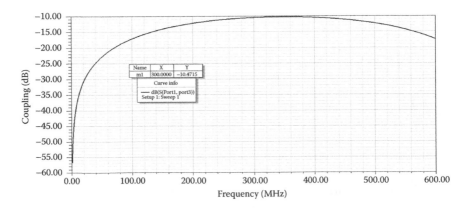

FIGURE 5.27   Simulated coupling level of two-layer 10 dB microstrip symmetrical directional coupler when the thickness of the first layer is 90 mil and the second layer is 60 mil.

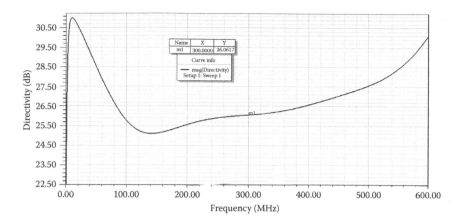

FIGURE 5.28 Simulated directivity level of two-layer 10 dB microstrip symmetrical directional coupler when the thickness of the first layer is 90 mil and the second layer is 60 mil.

is shown in Figure 5.28. The directivity is found to be 26.06 dB at 300 MHz for the modified coupler. This shows 3.8 dB improvement versus single-layer two-line symmetrical directional coupler.

Besides from the improvement in the level of directivity at the operational frequency when two-layer configuration is used, the directivity slope is lowered and hence the improved directivity is observed between 200 and 600 MHz.

The improvement in the directivity level is more clear when the directivity levels of single- and two-layer configurations are compared as illustrated in Figure 5.29.

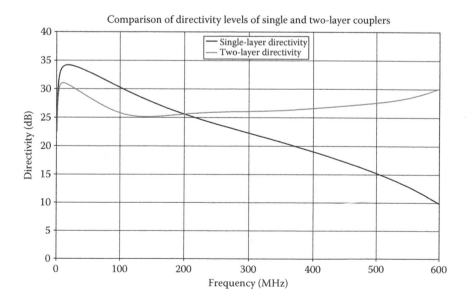

FIGURE 5.29 Comparison of the simulated directivity level for single- and two-layer 10 dB microstrip symmetrical directional coupler.

5.4 TRANSFORMER DIRECTIONAL COUPLERS

Transformer directional couplers are used for the lower frequency range up to 100 MHz due to their low cost in RF applications. Consider the transformer directional coupler shown in Figure 5.30. T_1 is the transformer with turns ratio $N_1{:}1$ and T_2 is the transformer with turns ratio $N_2{:}1$. The transformers are assumed to be ideal and lossless. The relations between voltages and currents through turns ratios of the directional coupler at the ports can be obtained as

$$V_2 = N_2(V_4 - V_3) \tag{5.57}$$

$$V_4 = N_1(V_2 - V_1) \tag{5.58}$$

and

$$I_1 = N_1(I_3 + I_4) \tag{5.59}$$

$$I_3 = N_1(I_1 + I_2) \tag{5.60}$$

Now, scattering parameters of the coupler can be obtained by using the incident and reflected waves, which are designated by a_i and b_i. Then, the voltages can be expressed in terms of waves as

$$V_i = \sqrt{Z}(a_i + b_i) \tag{5.61}$$

$$I_i = 1/\sqrt{Z}(a_i - b_i) \tag{5.62}$$

Z is the characteristic impedance at the ports of the directional coupler. The scattering parameters of the coupler is obtained by relating the incident and reflected waves as

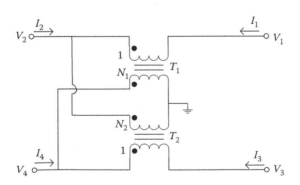

FIGURE 5.30 Transformer directional coupler.

$$S_{11} = -S_{44} = \frac{(-N_1^2 + N_2^2 - 2N_1N_2 + 1)}{(4N_1^2N_2^2 + 1 + (N_1^2 - N_2^2))} \tag{5.63}$$

$$S_{22} = -S_{33} = \frac{(-N_1^2 + N_2^2 + 2N_1N_2 - 1)}{(4N_1^2N_2^2 + 1 + (N_1^2 - N_2^2))} \tag{5.64}$$

$$S_{12} = S_{34} = \frac{(2N_1N_2)(2N_1N_2 - 1)}{(4N_1^2N_2^2 + 1 + (N_1^2 - N_2^2))} \tag{5.65}$$

$$S_{13} = -S_{24} = \frac{(-2N_1N_2)(N_1 + N_2)}{(4N_1^2N_2^2 + 1 + (N_1^2 - N_2^2))} \tag{5.66}$$

$$S_{14} = \frac{(-2N_1)(-N_1N_2 + N_2^2 + 1)}{(4N_1^2N_2^2 + 1 + (N_1^2 - N_2^2))} \tag{5.67}$$

$$S_{23} = \frac{(2N_2)(N_1N_2 - N_1^2 - 1)}{(4N_1^2N_2^2 + 1 + (N_1^2 - N_2^2))} \tag{5.68}$$

When the coupler is excited form port 1 and all other ports are matched, the coupling level is found from Equation 5.66 as

$$\text{Coupling level (dB)} = 20\log(-S_{13}) \tag{5.69}$$

The isolation and directivity levels are given by equations

$$\text{Isolation level (dB)} = 20\log(-S_{14}) \tag{5.70}$$

$$\text{Directivity level (dB)} = \text{coupling level (dB)} - \text{isolation level (dB)} \tag{5.71}$$

As a result, if only the number of turns are known, the amount of coupling, isolation, and directivity levels are found from Equations 5.63 through 5.71. However, there might be cases when the level of the voltage or the amount of current flowing through the coupler at certain points need to be known by the designer.

This can be accomplished by investigating the coupler shown in Figure 5.30 with circuit analysis techniques. The coupler shown in Figure 5.30 is modified for circuit analysis and voltage and current designations are shown in Figure 5.31. The application of Kirchhoff's current law (KCL) at node a and b in Figure 5.31 are

$$-I_1 + I_2 + I_s = 0 \tag{5.72a}$$

$$-I_3 + \frac{I_1}{N_1} - I_4 = 0 \tag{5.72b}$$

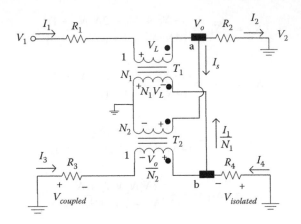

FIGURE 5.31   Transformer directional coupler for circuit analysis.

where

$$I_2 = \frac{V_o - V_2}{R_2} \tag{5.73}$$

$$I_s = \frac{V_3 + N_1 V_L}{R_3 N_2} + \frac{V_o}{R_3 N_2^2} \tag{5.74}$$

$$I_4 = \frac{N_1 V_L}{R_4} + \frac{V_4}{R_4} \tag{5.75}$$

$$I_3 = \frac{V_3}{R_3} + \frac{V_o}{N_2 R_3} + \frac{N_1 V_L}{R_3} \tag{5.76}$$

The analysis is performed when the coupler is working in the forward mode, that is

$$V_2 = V_3 = V_4 = 0 \tag{5.77}$$

Equations 5.73 through 5.77 reduce to

$$I_2 = \frac{V_o}{R_2} \tag{5.78}$$

$$I_3 = \frac{V_o}{N_2 R_3} + \frac{N_1 V_L}{R_3} \tag{5.79}$$

$$I_4 = \frac{N_1 V_L}{R_4} \tag{5.80}$$

$$I_s = \frac{N_1 V_L}{R_3 N_2} + \frac{V_o}{R_3 N_2^2} \tag{5.81}$$

The application of KVL on the upper portion of the coupler gives

$$V_L = V_1 - I_1 R_1 - V_o \tag{5.82}$$

Substitution of Equation 5.82 into Equation 5.72 gives

$$V_1 N_1 \left(\frac{1}{R_3} + \frac{1}{R_4} \right) + V_o \left(\frac{1}{R_3 N_2} - \frac{N_1}{R_3} - \frac{N_1}{R_4} \right) = I_1 \left(\frac{1}{N_1} + \frac{R_1 N_1}{R_3} + \frac{R_1 N_1}{R_4} \right) \tag{5.82a}$$

$$\frac{N_1 V_L}{R_3 N_2} + V_o \left(\frac{1}{R_3 N_2^2} - \frac{N_1}{R_3 N_2} + \frac{1}{R_2} \right) = I_1 \left(\frac{R_3 N_2 + R_1 N_1}{R_3 N_2} \right) \tag{5.82b}$$

From Equation 5.82b

$$I_1 = \left(\frac{N_1}{R_3 N_2 + R_1 N_1} \right) V_1 + \left(\frac{R_2 - N_1 N_2 R_2 + R_3 N_2^2}{N_2 R_2 \left(R_3 N_2 + R_1 N_1 \right)} \right) V_o \tag{5.83}$$

Let

$$B = \left(\frac{N_1}{R_3 N_2 + R_1 N_1} \right) \tag{5.84}$$

$$C = \left(\frac{R_2 - N_1 N_2 R_2 + R_3 N_2^2}{N_2 R_2 \left(R_3 N_2 + R_1 N_1 \right)} \right) \tag{5.85}$$

$$D = \left(\frac{1}{N_1} + \frac{R_1 N_1}{R_3} + \frac{R_1 N_1}{R_4} \right) \tag{5.86}$$

$$E = \left(\frac{1}{R_3 N_2} - \frac{N_1}{R_3} - \frac{N_1}{R_4} \right) \tag{5.87}$$

$$F = \left(\frac{1}{R_3} + \frac{1}{R_4} \right) \tag{5.88}$$

We can rewrite Equations 5.83 and 5.82a as

$$I_1 = BV_1 + CV_o \tag{5.89}$$

$$V_1 N_1 F + V_o E = I_1 D \tag{5.90}$$

Equations 5.88 and 5.89 lead to the following equations:

$$V_o = \frac{BD - N_1 F}{E - CD} V_1 \tag{5.91}$$

$$I_1 = \frac{BE - N_1 FC}{E - CD} V_1 \tag{5.92}$$

Now, substituting Equations 5.91 and 5.92 into Equation 5.81 gives the output voltage in terms of input voltage as

$$V_L = V_1 - \left(\frac{BE - N_1 FC}{E - CD} \right) R_1 V_1 - \left(\frac{BD - N_1 F}{E - CD} \right) V_1 \tag{5.93}$$

or

$$V_L = \left(\frac{E - CD - BER_1 + N_1 FCR_1 - BD + N_1 F}{E - CD} \right) V_1 \tag{5.94}$$

From KVL at the bottom of the coupler

$$V_{coupled} = N_1 V_L + \frac{V_o}{N_2} \tag{5.95}$$

$$V_{isolated} = N_1 V_L \tag{5.96}$$

The equivalent impedances that is seen by the excitation port can be found from

$$R_{in} = \frac{V_L + V_o}{I_1} \tag{5.97}$$

The performance parameters for the coupler is obtained from

$$\text{Coupling level (dB)} = 20 \log \left(\frac{V_{coupled}}{(R_2/R_1 + R_2) V_1} \right) \tag{5.98}$$

$$\text{Isolation level (dB)} = 20 \log \left(\frac{V_{isolated}}{(R_2/R_1 + R_2) V_1} \right) \tag{5.99}$$

$$\text{Directivity level (dB)} = \text{coupling level (dB)} - \text{isolation level (dB)} \quad (5.100)$$

and the return loss at the excitation port is found from

$$\text{Return loss (dB)} = 20\log\left(\left|\frac{R_1 - R_{in}}{R_1 + R_{in}}\right|\right) \quad (5.101)$$

In the analysis introduced, the voltage and current values are assumed to be the peak values or the amplitudes of the signal. The power at the input and output of the transformer is then found from current and the equivalent impedances seen at the corresponding port by using

$$P_{i,o} = \frac{1}{2}(I_{1,2})^2 R_{in,2} \quad (5.102)$$

Design Example: Transformer Directional Coupler

Design, simulate, build, and measure a transformer directional coupler with 20 dB coupling and directivity better than 30 dB for high-power RF application at 27.12 MHz when the input voltage is $V_{in,peak} = 100$ [V]. The port impedances are matched and are given to be equal to $R = 50$ [Ω]. Compare your analytical results with simulation results in time domain and frequency domain. Also, use the scattering parameters and compare your results with the circuit analysis results.

SOLUTION

20 dB coupling can be obtained when the number of turns is equal to $N_1 = N_2 = 10$ using scattering parameters derived by Equations 5.65 and 5.68. MATLAB GUI has been developed to calculate several design critical parameters using the derived Equations 5.72 through 5.102 and scattering parameters given by Equations 5.63 through 5.71. The parameters obtained using scattering parameters are valid only when all the ports are matched as is the case for the considered design problem. If ports are not matched, then the design parameters obtained by circuit analysis in the MATLAB GUI are the values that should be used. The program performs the directional analysis coupler in the forward mode, and accepts the excitation voltage at port 1, port impedances, and the number of turns as inputs.

The program then calculates the coupling, directivity, isolation levels, and several other design critical parameters as the program output. The program also produces design curves for coupling, directivity, and isolation levels versus the number of turns. The screen shot of the program showing the calculated design values for the required coupling level is given in Figure 5.32. Output, coupled and isolated voltages, power at the input and output, currents at the excitation and output port, coupling, isolation, directivity, and return loss are calculated and displayed. The transformer directional coupler is then simulated by Ansoft Designer in the frequency domain

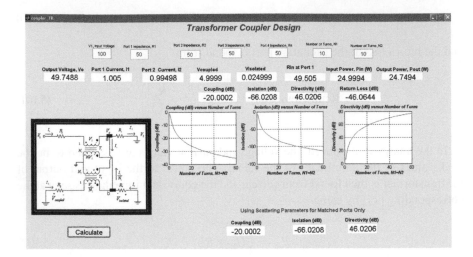

FIGURE 5.32   MATLAB GUI for the transformer coupler design.

using ideal transformers. This is a good practice to compare the analytical values using scattering parameters with the simulated values with the application of scattering parameters. The circuit that is simulated is shown in Figure 5.33.

The simulation results for coupling, isolation, and directivity are given by Figures 5.34 through 5.36. The coupling level and isolation are found to be −20 and −66.02 dB at the frequency of interest, respectively. Directivity is found to be 46.02 dB as shown in Figure 5.36. These values are in agreement with the values obtained by the program using the formulation in this section.

Since the calculated values and simulated values are in agreement, the transformer coupler can be built. However, the details about the transformer have to be determined. This includes the type of material that will be used as magnetic

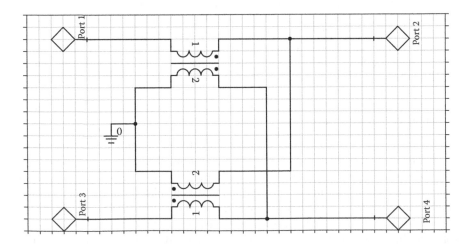

FIGURE 5.33   Frequency domain circuit simulator using S-parameters.

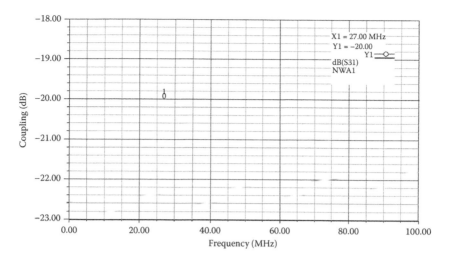

FIGURE 5.34 Simulated coupling level for the transformer coupler in the frequency domain.

core, winding information, inductance information, and so on. Hence, it is beneficial for the designer to use the calculated values of the transformer design in the time domain circuit simulator, which will take into account of coupling as it is the case in real-world application. The core material is chosen to be −7, which is carbonyl TH with a permeability of 9 and has good performance for applications when the frequency of operation is between 3 and 35 MHz. The core dimensions are found to be OD = 1.75 cm, ID = 0.94 cm, and h = 0.48 cm. This core is T-68-7 with white color code. Three cores are stacked and 16AWG wire is used for winding. Ten turns results in the transformer inductance value of 1.61 µH and this gives

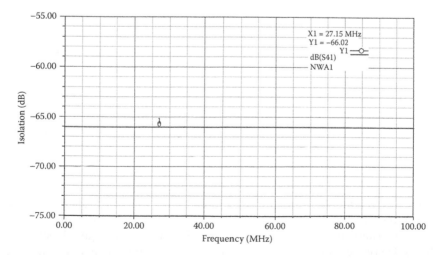

FIGURE 5.35 Simulated isolation level for the transformer coupler in the frequency domain.

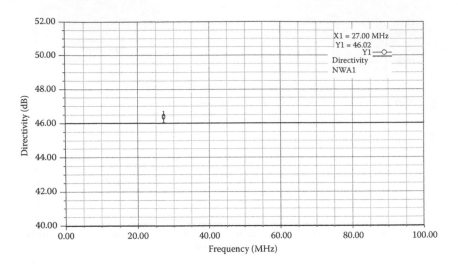

FIGURE 5.36 Simulated directivity level for the transformer coupler in the frequency domain.

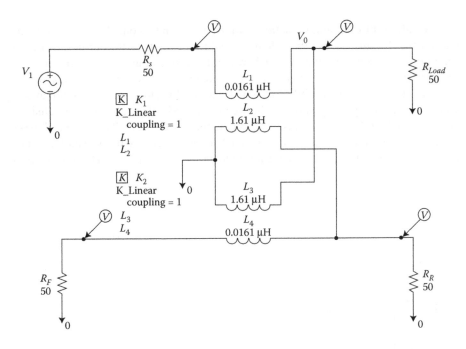

FIGURE 5.37 Time domain simulation of the transformer coupler.

an impedance value that is more than 5 times higher than the higher impedance termination, which is 50 Ω. Hence, the transformer design should produce correct operating conditions at the frequency of operation. Since, the required inductance value is determined, this value now can be used in the time domain circuit simulator to obtain all the parameters versus time and take into account of the actual approximate coupling. The circuit that is simulated by Pspice for the desired transformer coupler is shown in Figure 5.37.

Simulation results showing the coupling, isolation, and directivity are given by Figures 5.38 and 5.39. Coupling, isolation, and directivity in Pspice are found using macros in the probe section with the application of formulation given by Figure 5.40. The coupling level and isolation are found to be −20 and −65.87 dB, respectively. The directivity is found to be 45.87 dB based on the time domain simulation. Simulated values are in agreement with the frequency domain simulator and MATLAB GUI program. As a result, we can go ahead and build the coupler using the calculated values, which are confirmed by the time and frequency domain simulators. The constructed transformer coupler is shown in Figure 5.41. The single turn is accomplished by using coax semirigid cable with 50 Ω characteristic impedance, 0.25 in outer

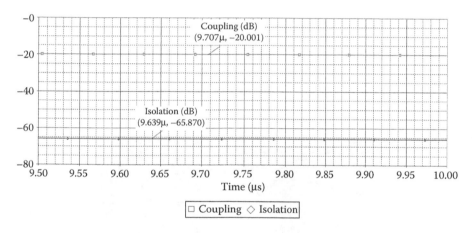

FIGURE 5.38 Simulated coupling and isolation levels for the transformer coupler in the time domain.

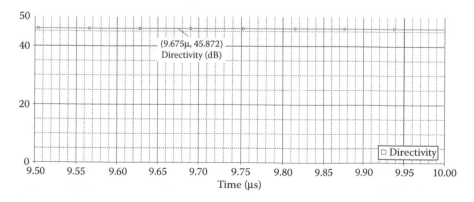

FIGURE 5.39 Simulated directivity level for the transformer coupler in the time domain.

Macros   ☒

Definition [   ]

FileName   [   ]

```
Coupling = avg(20*log10(V(Vcoupled)/((0.5)*(V(V1:+)))))
Directivity = Coupling-Isolation
Isolation = avg(20*log10(V(Visolated)/((0.5)*(V(V1:+)))))
pi = 3.14159265
```

| Save | Save To... | Delete | Load | Close |

FIGURE 5.40   Macros used in Pspice for coupling, isolation, and directivity simulation.

FIGURE 5.41   Constructed transformer coupler for 27.12 MHz operation.

conductor diameter, and 0.076 in inner conductor diameter. Polytetrafluoroethylene (PTFE) is used as a dielectric material with 0.214 in diameter. The length of the semirigid coax cable used in the construction of the coupler is 3.75 cm. Semirigid coax cable outline is illustrated in Figure 5.42. As it is planned, T-68-7 toroidal core is used with 10 turns and winding is done with 16AWG enameled wire. Three-mil-thick Teflon tape is used to the insulated rigid cable from the core and core

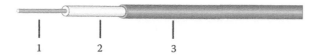

FIGURE 5.42 Semirigid coax cable used in the transformer coupler.

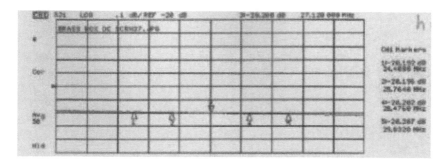

FIGURE 5.43 Measured coupling of the constructed transformer coupler.

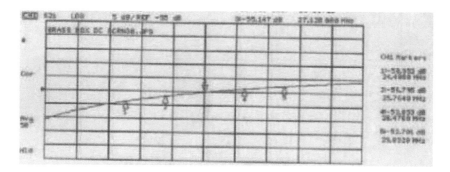

FIGURE 5.44 Measured isolation of the constructed transformer coupler.

windings. N-type connectors are used due to high-power application requirement. The measured frequency response of the coupler for coupling and isolation are given in Figures 5.43 and 5.44.

As seen, the measured coupling and isolation levels at 27.12 MHz are −20.2 and −55.147 dB. This gives the measured value of 34.947 dB directivity. The measured input impedance versus frequency is shown in Figure 5.45. It is seen that the input impedance, which is calculated as R_{in}, is equal to $(49.691 + j3.2676)\ \Omega$. The real part is in agreement with the calculated $R_{in} = 49.7488\ [\Omega]$. The reactance always exists because the used components are lossy. The measured inductive reactance can be eliminated by using the compensation capacitor with ease. The measured input impedance corresponds to VSWR of 1.07, which is an indication of a well-matched

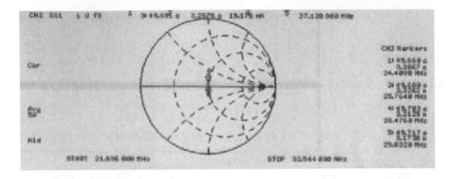

FIGURE 5.45   Measured input impedance of the constructed transformer coupler.

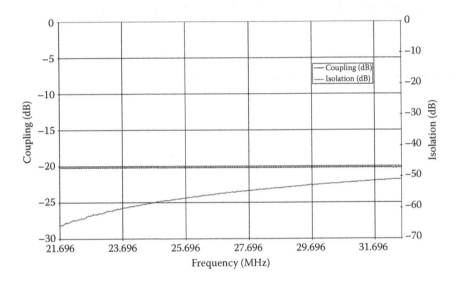

FIGURE 5.46   Measured isolation and coupling of the constructed transformer coupler.

system. This can be further improved with the implementation of the compensation capacitor. The data for the measured values are extracted for coupling and isolation from the network analyzer and Excel is used to visually see the results for comparison better as illustrated in Figure 5.46.

In addition, the directivity response of the coupler versus frequency is obtained by this way and shown in Figure 5.47. The measured values show that directivity and hence the isolation of the coupler gets worse as the frequency increases. The coupling level is maintained across the frequency bandwidth.

Overall, the designed system meets with the requirements outlined in the problem statement. It is shown that hybrid simulation of the coupler is crucial to design better, which meets the design requirements with a cost-effective way.

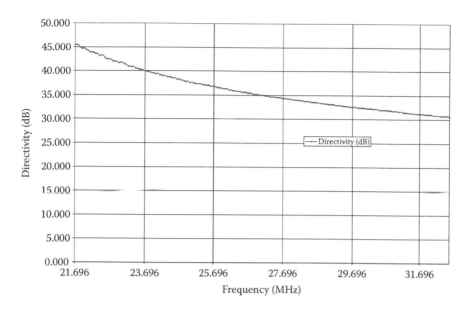

FIGURE 5.47 Measured directivity of the constructed transformer coupler.

REFERENCES

1. B.M. Oliver. Directional electromagnetic couplers, *Proceedings of IRE*, 42, 1686–1692, 1954.
2. G.D. Monteath. Coupled transmission lines as symmetrical directional couplers, *IEE Proceedings*, 102, part B, 383–392, 1955.
3. J. Reed and J. Wheeler. A method of analysis of symmetrical four port networks, *IRE Transactions on Microwave Theory and Techniques*, 4, 246–252, 1956.
4. T.G. Bryant and J.A. Weiss. Parameters of microstrip transmission lines and coupled pairs of microstrip lines, *IEEE Transactions on Microwave Theory and Techniques*, MTT-16(12), 1021–1027, 1968.
5. S. Akhtarzad, T.R. Rowbotham, and P.B. Jones. The design of coupled microstrip lines, *IEEE Transactions on Microwave Theory and Techniques*, MTT-23(7), 486–492, 1975.
6. K.C. Gupta, R. Garg, and R. Chadha. *Computer-Aided Design of Microwave Circuits*, Artech House, Massachusetts, Chapter 3, 1981.
7. M. Kirschning and R.H. Jansen. Accurate wide-range design equations for frequency-dependent characteristic of parallel coupled microstrip lines, *IEEE Transactions on Microwave Theory and Techniques*, MTT-32(1), 83–90, 1984.
8. J.J. Hinton. On design of coupled microstrip lines, *IEEE Transactions on Microwave Theory and Techniques*, MTT-28(3), 272, 1980.
9. T. Emery, Y. Chin, H. Lee, and V.K. Tripathi. Analysis and design of ideal non symmetrical coupled microstrip directional couplers, *IEEE MTT-S Symposium Digest*, 1, 329–332, 1989.
10. C. Tsai and K.C. Gupta. A generalized model for coupled lines and its applications to two-layer planar circuits, *IEEE Transactions on Microwave Theory and Techniques*, 40, 2190–2199, 1992.
11. S. Gruszczynski and K. Wincza. Generalized methods for the design of quasi-ideal symmetric and asymmetric coupled-line sections and directional couplers, *IEEE Transactions on Microwave Theory and Techniques*, 59(7), 1709–1718, 2011.

REFERENCES

6 MF-UHF Filter Design Techniques

6.1 INTRODUCTION

Filters are one of the indispensable components in RF applications [1]. An ideal filter provides the perfect transmission of the signal for frequencies in the required pass-band region, and the infinite attenuation for frequencies in the stopband region. The critical RF filter parameters include high return loss, minimum attenuation distortion, a flat group delay in the passband, and high attenuation in the stopband. The basic filters that are used in RF applications are categorized as low-pass, high-pass, bandpass, and bandstop filters with their ideal filter characteristics shown in Figure 6.1.

6.2 FILTER DESIGN PROCEDURE

Filters can be analyzed as lossless linear two-port networks using network parameters [2–12]. The conventional filter design procedure for low-pass, high-pass, bandpass, or bandstop filters begins from low-pass filter (LPF) prototype and then involves imped-ance and frequency scaling, and filter transformation to high-pass, bandpass, or band-stop filters to obtain the final component values at the frequency of operation. The

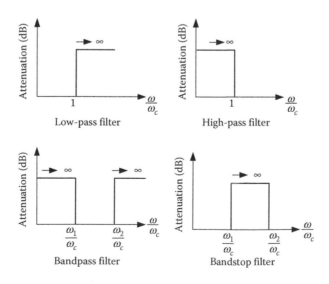

FIGURE 6.1   Ideal filter characteristics.

design is then simulated and compared with specifications. The final step in the design of filters involves the implementation and measurement of the filter response. This design procedure can be shown as a block diagram illustrated in Figure 6.2. The LPF prototype is the basic building block in the filter design. The attenuation profile of the LPF is then a critical parameter in the design procedure. The attenuation profiles of the LPF can be binomial (Butterworth), Chebyshev, or elliptic (Cauer) as shown in Figure 6.3. Binomial filters provide monotonic attenuation profile and need more components to achieve steep attenuation transition from passband to stopband whereas Chebyshev filters have steeper slope and equal amplitude ripples in the passband.

Elliptic filters have steeper transition from passband to stopband similar to Chebyshev filters and exhibit equal amplitude ripples in the passband and stopband. RF/microwave filters and filter components can be represented using a two-port network shown in Figure 6.4. The network analysis can be conducted using *ABCD*

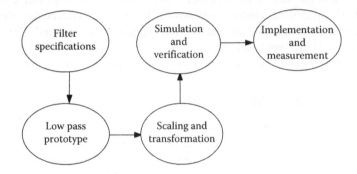

FIGURE 6.2   Filter design block diagram.

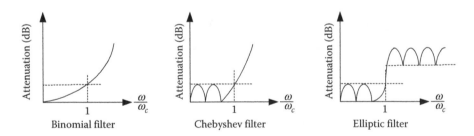

FIGURE 6.3   Attenuation profiles of low-pass filter.

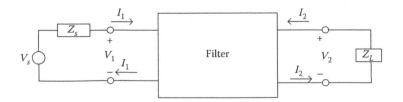

FIGURE 6.4   Two-port network representation.

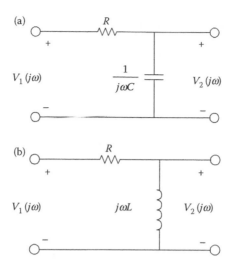

FIGURE 6.5 Transfer function analysis circuits. (a) Low-pass filter. (b) High-pass filter.

parameters for each filter. The filter elements can be considered as cascaded components and hence the overall *ABCD* parameter of the network is just simple matrix multiplication of the *ABCD* parameter for each element. The characteristics of the filter in practice is determined via insertion loss, S_{21}, and return loss, S_{11}. *ABCD* parameters can be converted to scattering parameters and insertion loss and return loss for the filter can be determined. The insertion loss of filter networks shown in Figures 6.5a and 6.5b will be analyzed using *ABCD* parameters and the cut-off frequency for each network will be determined using transfer functions. Low-pass filter and high-pass filter (HPF) circuits for transfer function derivation to obtain cut-off frequency are given by Figures 6.5a and 6.5b, respectively.

The cut-off frequency of the circuit for the LPF in Figure 6.5a is determined using the transfer function given by

$$H(j\omega) = \frac{V_2}{V_1} = \frac{1/RC}{j\omega + 1/RC} \tag{6.1}$$

Cut-off frequency occurs when

$$|H(j\omega_c)| = \frac{1}{\sqrt{2}} H_{max} \tag{6.2}$$

From Equation 6.1

$$|H(j\omega)| = \frac{1/RC}{\sqrt{\omega^2 + (1/RC)^2}} \tag{6.3}$$

The maximum value of the transfer function is found as

$$H_{\max} = |H(j0)| = 1 \tag{6.4}$$

Using Equations 6.3 and 6.4, we have

$$|H(j\omega_c)| = \frac{1}{\sqrt{2}}(1) = \frac{1/RC}{\sqrt{\omega_c^2 + (1/RC)^2}} \tag{6.5}$$

Equation 6.5 is satisfied when

$$\omega_c = \frac{1}{RC} \tag{6.6}$$

The cut-off frequency is found from Equation 6.6 as

$$f_c = \frac{1}{2\pi RC} \tag{6.7}$$

The same analysis can be conducted for the HPF circuit shown in Figure 6.5b. The transfer function of the circuit is obtained as

$$H(j\omega) = \frac{j\omega}{j\omega + (R/L)} \tag{6.8}$$

From Equation 6.8

$$|H(j\omega)| = \frac{\omega}{\sqrt{\omega^2 + (R/L)^2}} \tag{6.9}$$

and

$$H_{\max} = |H(j\infty)| = 1 \tag{6.10}$$

Then, the cut-off frequency for the HPF is found from

$$\frac{1}{\sqrt{2}} = |H(j\omega_c)| = \frac{\omega_c}{\sqrt{\omega_c^2 + (R/L)^2}} \tag{6.11}$$

which gives the cut-off frequency as

$$\omega_c = \frac{R}{L} \quad \text{or} \quad f_c = \frac{1}{2\pi}\left(\frac{R}{L}\right) \tag{6.12}$$

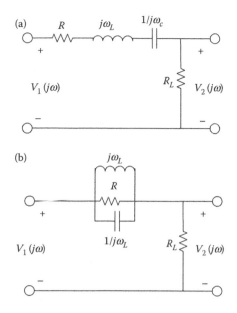

FIGURE 6.6   Transfer function analysis circuits. (a) Bandpass filter. (b) Bandstop filter.

The resonant frequency and the cut-off frequencies of the bandpass filter (BPF) shown in Figure 6.6a are analyzed using the transfer function

$$H(j\omega) = \frac{R_L/L \; j\omega}{(j\omega)^2 + ((R_L + R)/L) \, j\omega + (1/LC)} \qquad (6.13)$$

The magnitude of the transfer function is

$$|H(j\omega)| = \frac{R_L/L\omega}{\sqrt{((1/LC) - \omega^2)^2 + (\omega((R_L + R)/L))^2}} \qquad (6.14)$$

The maximum value of the transfer function occurs at the resonant frequency f_o and is equal to

$$H_{max} = |H(j\omega_o)| = \frac{R}{R + R_i} \qquad (6.15)$$

where

$$\omega_o = \frac{1}{\sqrt{LC}} \quad \text{or} \quad f_o = \frac{1}{2\pi\sqrt{LC}} \qquad (6.16)$$

The cut-off frequencies for the BPF are found as

$$\omega_{c1} = -\frac{R_L + R}{2L} + \sqrt{\left(\frac{R_L + R}{2L}\right)^2 + \frac{1}{LC}} \qquad (6.17a)$$

$$\omega_{c2} = \frac{R_L + R}{2L} + \sqrt{\left(\frac{R_L + R}{2L}\right)^2 + \frac{1}{LC}} \qquad (6.17b)$$

and it can be shown that

$$\omega_o = \sqrt{\omega_{c1} \cdot \omega_{c2}} \qquad (6.18)$$

Bandstop filter (BSF) can be analyzed similarly by replacing the series RLC circuit with parallel RLC circuit as shown in Figure 6.6b.

Low-pass and high-pass filters have four sections and are cascaded as shown in Figure 6.7. The insertion loss using the *ABCD* parameter of the LPF in Figure 6.7a can be obtained from

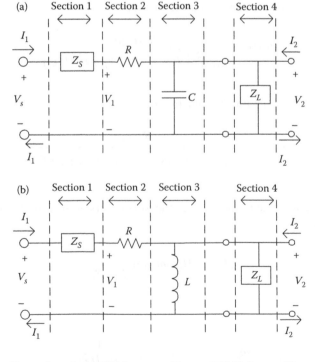

FIGURE 6.7   Network analysis of (a) low-pass filter and (b) high-pass filter.

$$\begin{bmatrix} A & B \\ C & D \end{bmatrix} = \begin{bmatrix} 1 & Z_S \\ 0 & 1 \end{bmatrix} \begin{bmatrix} 1 & R \\ 0 & 1 \end{bmatrix} \begin{bmatrix} 1 & 0 \\ j\omega C & 1 \end{bmatrix} \begin{bmatrix} 1 & 0 \\ \dfrac{1}{Z_L} & 1 \end{bmatrix} \tag{6.19a}$$

or

$$\begin{bmatrix} A & B \\ C & D \end{bmatrix} = \begin{bmatrix} 1 + (R + Z_S)\left(j\omega C + \dfrac{1}{Z_L}\right) & R + Z_S \\ j\omega C + \dfrac{1}{Z_L} & 1 \end{bmatrix} \tag{6.19b}$$

The insertion loss is then found from

$$\text{Insertion loss (dB)} = 20 \log (|S_{21}|) \tag{6.20}$$

S_{21} for the LPF shown in Figure 6.7a when $Z_S = Z_L = Z_o$ is

$$S_{21} = \frac{2V_2}{V_S} = \frac{2}{A} = \frac{2}{1 + (R + Z_o)(j\omega C + (1/Z_o))} \tag{6.21}$$

The response is obtained by MATLAB® using the formulation given by Equations 6.20 and 6.21 and shown in Figure 6.8 for $C = 8$ pF, $R = 100\ \Omega$, and $Z_o = 50\ \Omega$.

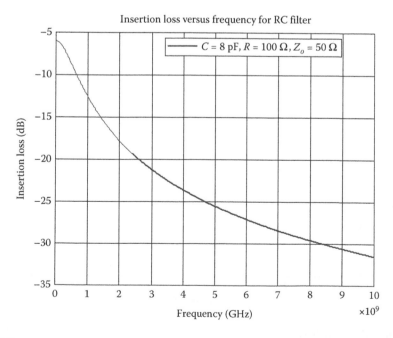

FIGURE 6.8 Insertion loss for low-pass filter when $C = 8$ pF, $R = 100\ \Omega$, and $Z_o = 50\ \Omega$.

The LPF circuit is simulated with the frequency domain circuit simulator and the results are compared with the results obtained with MATLAB. Figure 6.9 is the LPF that is simulated and Figure 6.10 is the response versus frequency for the insertion loss.

ABCD parameters for the HPF network shown in Figure 6.7b can be written as

$$\begin{bmatrix} A & B \\ C & D \end{bmatrix} = \begin{bmatrix} 1 & Z_S \\ 0 & 1 \end{bmatrix}\begin{bmatrix} 1 & R \\ 0 & 1 \end{bmatrix}\begin{bmatrix} 1 & 0 \\ \dfrac{1}{j\omega L} & 1 \end{bmatrix}\begin{bmatrix} 1 & 0 \\ \dfrac{1}{Z_L} & 1 \end{bmatrix} \tag{6.22}$$

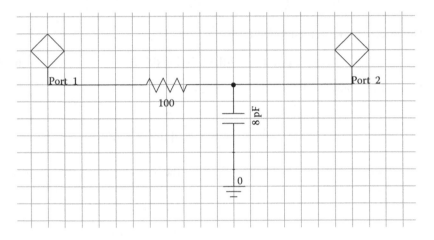

FIGURE 6.9   Low-pass filter simulation when $C = 8$ pF, $R = 100\ \Omega$, and $Z_o = 50\ \Omega$.

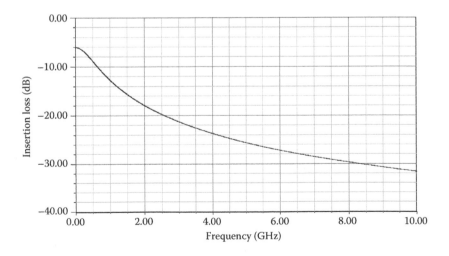

FIGURE 6.10   Simulated insertion loss for low-pass filter when $C = 8$ pF, $R = 100\ \Omega$, and $Z_o = 50\ \Omega$.

or

$$\begin{bmatrix} A & B \\ C & D \end{bmatrix} = \begin{bmatrix} 1 + (R + Z_S)\left(\dfrac{1}{j\omega L} + \dfrac{1}{Z_L} \right) & R + Z_S \\ \dfrac{1}{j\omega L} + \dfrac{1}{Z_L} & 1 \end{bmatrix}$$

(6.23)

S_{21} for the HPF shown in Figure 6.7b when $Z_S = Z_L = Z_o$ is

$$S_{21} = \frac{2V_2}{V_S} = \frac{2}{A} = \frac{2}{1 + (R + Z_o)((1/j\omega L) + (1/Z_o))}$$

(6.24)

The insertion loss is found from Equation 6.20. The response is obtained by MATLAB using the formulation given by Equations 6.20 and 6.21 and shown in Figure 6.11 for $L = 5$ nH, $R = 5$ Ω, and $Z_o = 50$ Ω. The results are confirmed with the circuit simulator. Bandpass and bandstop filter networks can be analyzed using the networks shown in Figures 6.12a and b, respectively. Both filters have three sections where section 2 is a series RLC circuit for BPF and a parallel RLC circuit for BSF.

The $ABCD$ parameter of the BPF network circuit in Figure 6.12a is

$$\begin{bmatrix} A & B \\ C & D \end{bmatrix} = \begin{bmatrix} 1 & Z_S \\ 0 & 1 \end{bmatrix} \begin{bmatrix} 1 & R + j\omega L + (1/j\omega C) \\ 0 & 1 \end{bmatrix} \begin{bmatrix} 1 & 0 \\ \dfrac{1}{Z_L} & 1 \end{bmatrix}$$

(6.25)

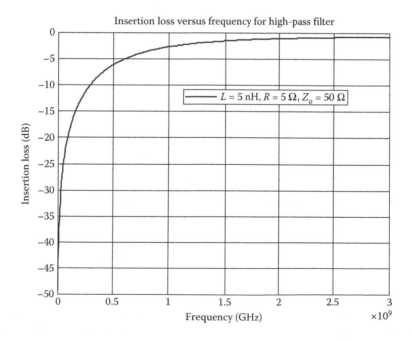

FIGURE 6.11 Insertion loss for high-pass filter when $L = 5$ nH, $R = 5$ Ω, and $Z_o = 50$ Ω.

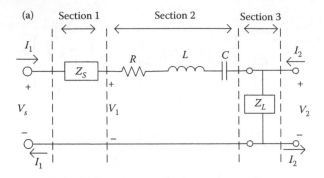

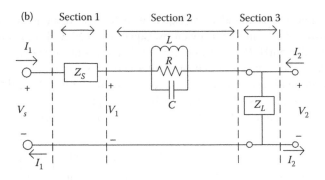

FIGURE 6.12   Network analysis of (a) bandpass filter and (b) bandstop filter.

or

$$
\begin{bmatrix} A & B \\ C & D \end{bmatrix} = \begin{bmatrix} 1 + \dfrac{R + j\omega L + (1/j\omega C) + Z_s}{Z_L} & R + j\omega L + (1/j\omega C) + Z_s \\[4mm] \dfrac{1}{Z_L} & 1 \end{bmatrix}
\tag{6.26}
$$

S_{21} for the BPF shown in Figure 6.12a when $Z_S = Z_L = Z_o$ is

$$
S_{21} = \frac{2V_2}{V_S} = \frac{2}{A} = \frac{2}{1 + \dfrac{R + j\omega L + (1/j\omega C) + Z_s}{Z_L}}
\tag{6.27}
$$

The insertion loss is found from Equation 6.20. The response for insertion loss is obtained by MATLAB using the formulation given by Equations 6.20 and 6.27 as shown in Figure 6.13.

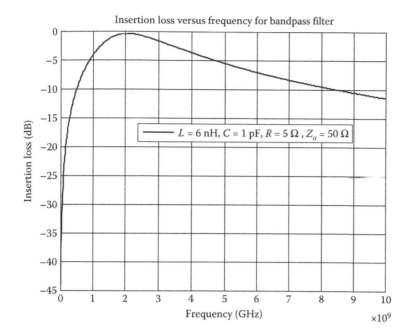

FIGURE 6.13   Insertion loss for bandpass filter when $L = 6$ nH, $C = 1$ pF, $R = 5$ Ω, and $Z_o = 50$ Ω.

The resonant frequency for the BPF from Equation 6.16 is

$$f_o = \frac{1}{2\pi\sqrt{LC}} = \frac{1}{2\pi\sqrt{(6 \times 10^{-9})(1 \times 10^{-12})}} = 2.054 \text{ [GHz]} \quad (6.28)$$

as confirmed with MATLAB results shown in Figure 6.13. The *ABCD* parameter of the BSF network circuit in Figure 6.12b can be derived similarly as

$$\begin{bmatrix} A & B \\ C & D \end{bmatrix} = \begin{bmatrix} 1 & Z_S \\ 0 & 1 \end{bmatrix} \begin{bmatrix} 1 & \dfrac{1}{G + j\omega C + (1/j\omega L)} \\ 0 & 1 \end{bmatrix} \begin{bmatrix} 1 & 0 \\ \dfrac{1}{Z_L} & 1 \end{bmatrix} \quad (6.29)$$

or

$$\begin{bmatrix} A & B \\ C & D \end{bmatrix} = \begin{bmatrix} 1 + \dfrac{\dfrac{1}{G + j\omega C + (1/j\omega L)} + Z_s}{Z_L} & \dfrac{1}{G + j\omega C + (1/j\omega L)} + Z_s \\ \dfrac{1}{Z_L} & 1 \end{bmatrix} \quad (6.30)$$

S_{21} for BSF shown in Figure 6.12b when $Z_S = Z_L = Z_o$ is

$$S_{21} = \frac{2V_2}{V_S} = \frac{2}{A} = \frac{2}{1 + \dfrac{1}{G + j\omega C + (1/j\omega L)} + Z_s/Z_L} \tag{6.31}$$

The insertion loss is found from Equation 6.20. The response for the insertion loss is obtained by MATLAB using the formulation given by Equations 6.20 and 6.31 as shown in Figure 6.14. The resonant frequency for the BSF is again

$$f_o = \frac{1}{2\pi\sqrt{LC}} = \frac{1}{2\pi\sqrt{(2 \times 10^{-9})(3 \times 10^{-12})}} = 2.054 \ [\text{GHz}] \tag{6.32}$$

The return loss for any filter discussed is obtained using the relation

$$S_{11} = \frac{Z_{in} - Z_o}{Z_{in} + Z_o} \tag{6.33}$$

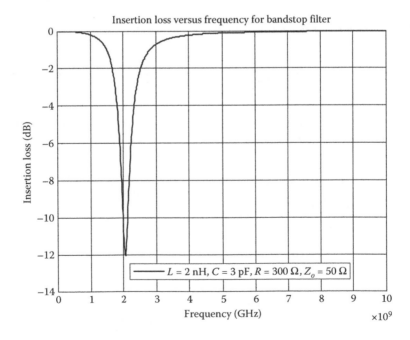

FIGURE 6.14 Insertion loss for bandstop filter when $L = 2$ nH, $C = 3$ pF, $R = 300 \ \Omega$, and $Z_o = 50 \ \Omega$.

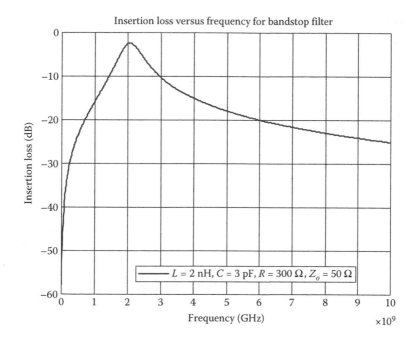

FIGURE 6.15 Return loss for bandstop filter when $L = 2$ nH, $C = 3$ pF, $R = 300$ Ω, and $Z_o = 50$ Ω.

For BSF

$$Z_{in} = \frac{1}{G + j\omega C + (1/j\omega L)} + Z_s \qquad (6.34)$$

The return loss is calculated from

$$\text{Return loss (dB)} = 20 \log (|S_{11}|) \qquad (6.35)$$

The return loss of the BSF with $L = 2$ nH, $C = 3$ pF, $R = 300$ Ω, and $Z_o = 50$ Ω is obtained versus frequency and is given in Figure 6.15.

6.3 FILTER DESIGN BY INSERTION LOSS METHOD

There are mainly two filter synthesis methods in the design of RF filters: image parameter method and insertion loss method. Although the design procedure with the image parameter method is straightforward and easy, it is not possible to realize an arbitrary frequency response with the use of this method. We will be applying the insertion loss method to design and implement the filters in this section. The filter design with the insertion loss method begins with complete filter specifications as shown in Figure 6.2. Filter specifications are used to identify the prototype filter values and the prototype filter circuit is synthesized. Scaling and transformation of the prototype values are performed to have the final filter component values. The prototype element values of the LPF circuit are obtained using the power loss ratio.

6.3.1 Low-Pass Filters

Consider the two-element LPF prototype shown in Figure 6.16.

In the insertion loss method, the filter response is defined by the power loss ratio, P_{LR}, which is defined by

$$P_{LR} = \frac{P_{incident}}{P_{load}} = \frac{1}{1 - |S_{11}|^2} \tag{6.36}$$

where

$$S_{11} = \frac{Z_{in} - 1}{Z_{in} + 1} \tag{6.37}$$

$P_{incident}$ refers to the available power from the source and P_{load} represents the power delivered to the load. As explained earlier, attenuation characteristics of the filter fall into one of these three categories: binomial (Butterworth), Chebyshev, or elliptic. The binomial or Butterworth response provides the flattest passband response for a given filter and is defined by

$$P_{LR} = 1 + k^2 \left(\frac{\omega}{\omega_c} \right)^{2N} \tag{6.38}$$

where $k = 1$ and N is the order of the filter. Chebyshev filters provide steeper transition from passband to stopband while they have equal ripples in the passband and their attenuation characteristics are defined by

$$P_{LR} = 1 + k^2 T_N^2 \left(\frac{\omega}{\omega_c} \right) \tag{6.39}$$

where T_N is the Chebyshev polynomial. The prototype values of the filter circuit, L and C, shown in Figure 6.16 are found by solving the equations given in Equations 6.36 through 6.39.

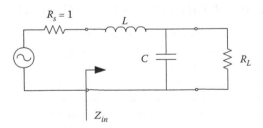

FIGURE 6.16 Two-element low-pass prototype circuit.

6.3.1.1 Binomial Filter Response

Binomial filter response can be explained using Figure 6.16 for two-element LPF network. The source impedance and cut-off frequency for the circuit in Figure 6.16 are assumed to be 1 Ω and 1 rad/s, respectively. In the LPF network, $N = 2$ and the power loss becomes

$$P_{LR} = 1 + \omega^4 \tag{6.40}$$

The input impedance is found as

$$Z_{in} = \frac{j\omega L(1 + \omega^2 R^2 C^2) + R(1 - j\omega RC)}{\left(1 + \omega^2 R^2 C^2\right)} \tag{6.41}$$

which leads to

$$S_{11} = \frac{[(j\omega L(1 + \omega^2 R^2 C^2) + R(1 - j\omega RC))/(1 + \omega^2 R^2 C^2)] - 1}{[(j\omega L(1 + \omega^2 R^2 C^2) + R(1 - j\omega RC))/(1 + \omega^2 R^2 C^2)] + 1} \tag{6.42}$$

When Equation 6.42 is substituted into Equation 6.36 with 6.40, the only L and C values satisfying the equation are found to be

$$L = C = 1.4142 \tag{6.43}$$

The same procedure is applied for LPF circuit with any number. The values obtained with this method are tabulated in Table 6.1. In essence, the two-element low-pass proto circuit we analyzed is called as a ladder network. Although the LPF circuit in Figure 6.16 begins with the series inductor, our analysis applies when it is switched with the shunt capacitor. In addition, the number of elements can be increased to N and the ladder network can be generalized as shown in Figure 6.17 where the values shown in Table 6.1 can be used for binomial response.

TABLE 6.1

Component Values for Binomial Low-Pass Filter Response with $g_0 = 1$, $\omega_c = 1$

N	g_1	g_2	g_3	g_4	g_5	g_6	g_7	g_8	g_9
1	2.0000	1.0000							
2	1.4142	1.4142	1.0000						
3	1.0000	2.0000	1.0000	1.0000					
4	0.7654	1.8478	1.8478	0.7654	1.0000				
5	0.6180	1.6180	2.0000	1.6180	0.6180	1.0000			
6	0.5176	1.4142	1.9318	1.9318	1.4142	0.5176	1.0000		
7	0.4450	1.2470	1.8019	2.0000	1.8019	1.2470	0.4450	1.0000	
8	0.3902	1.1111	1.6629	1.9615	1.9615	1.6629	1.1111	0.3902	1.0000

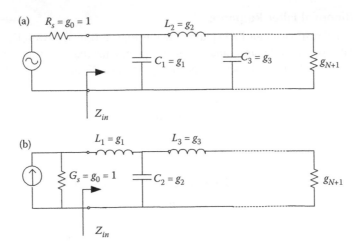

FIGURE 6.17 Low-pass prototype ladder networks. (a) First-element shunt C. (b) First-element series L.

In the low-pass prototype circuits in Figure 6.17, g_0 represents the source resistance or conductance, whereas g_{N+1} represents the load resistance or conductance. g_N is an inductor for the series-connected component and a capacitor for the parallel-connected component. The attenuation curves for the low-pass prototype filters can be found from Equations 6.37 and 6.39 as

$$\text{Attenuation (dB)} = 10 \log (P_{LR}) \tag{6.44}$$

The attenuation curves for the binomial response using Equation 6.44 is obtained by MATLAB and given in Equation 6.18. Once the design filter specifications are given, the number of required elements to have the desired attenuation is determined from the attenuation curves. In the second step, the table is used to determine the normalized component values for the required number of elements found in the previous stage. Then, the scaling and transformation step is performed and the final filter component values are obtained.

Since the original normalized component values of the LPF filter are designated as L, C, and R_L, the final scaled component values of the filter with source impedance R_o is found from

$$R'_s = R_o \tag{6.45}$$

$$R'_L = R_o R_L \tag{6.46}$$

$$L' = R_o \frac{L_n}{\omega_c} \tag{6.47}$$

$$C' = \frac{C_n}{R_o \omega_c} \tag{6.48}$$

Example 1: Maximally Flat Low-Pass Filter Response

Design a binomial LPF with cut-off frequency $f_c = 1.5$ GHz to have a minimum 35 dB attenuation at 4.5 GHz when the source and load impedances are 50 Ω.

SOLUTION

Step 1—Use Figure 6.18 to determine the required number of elements to obtain a minimum 35 dB attenuation at $f = 4.5$ GHz.

$$\left| \frac{\omega}{\omega_c} \right| - 1 = \frac{4.5}{1.5} - 1 = 2 \rightarrow N = 4$$

Step 2—Use Table 6.1 to determine the normalized LPF component values. We choose to begin with the series inductor as shown in Figure 6.19. The normalized component values for the filter are obtained from the table as

$$L_1 = g_1 = 0.7654, \quad L_2 = g_3 = 1.8478, \quad C_1 = g_2 = 1.8478$$

Step 3—Apply the impedance and frequency scaling

$$R_s' = R_o = 50 \ [\Omega]$$

$$R_L' = R_o R_L = 50(1) = 50 \ [\Omega]$$

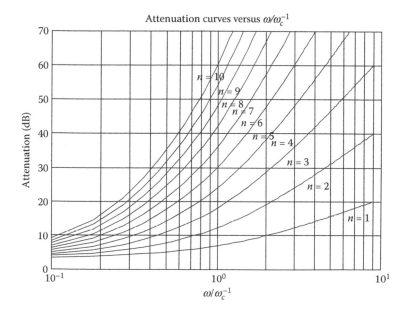

FIGURE 6.18 Attenuation curves for binomial filter response for low-pass prototype circuits.

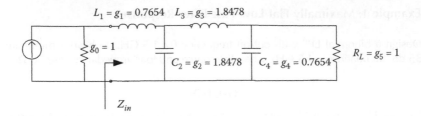

FIGURE 6.19 Fourth-order normalized LPF for binomial response.

$$L_1' = R_o \frac{L_n}{\omega_c} = 50 \frac{0.7654}{(2\pi \times 1.5 \times 10^9)} = 4.06 \text{ [nH]}$$

$$C_2' = \frac{C_n}{R_o \omega_c} = \frac{1.8478}{50(2\pi \times 1.5 \times 10^9)} = 3.92 \text{ [pF]}$$

$$L_3' = R_o \frac{L_n}{\omega_c} = 50 \frac{1.8478}{(2\pi \times 1.5 \times 10^9)} = 9.8 \text{ [nH]}$$

$$C_4' = \frac{C_n}{R_o \omega_c} = \frac{0.7654}{50(2\pi \times 1.5 \times 10^9)} = 1.62 \text{ [pF]}$$

The final LPF circuit having binomial filter response is shown in Figure 6.20. The final circuit shown in Figure 6.20 is analyzed using network parameters and the insertion loss is obtained with *ABCD* parameters for the cascaded components as previously discussed.

$$\begin{bmatrix} A & B \\ C & D \end{bmatrix} = \begin{bmatrix} 1 & Z_S \\ 0 & 1 \end{bmatrix} \begin{bmatrix} 1 & j\omega L_1' \\ 0 & 1 \end{bmatrix} \begin{bmatrix} 1 & 0 \\ j\omega C_2' & 1 \end{bmatrix} \begin{bmatrix} 1 & j\omega L_3' \\ 0 & 1 \end{bmatrix} \begin{bmatrix} 1 & 0 \\ j\omega C_4' & 1 \end{bmatrix} \begin{bmatrix} 1 & 0 \\ \dfrac{1}{Z_L} & 1 \end{bmatrix} \quad (6.49)$$

The insertion loss is obtained from Equations 6.20 and 6.21. MATLAB is used to plot the insertion loss from Equation 6.49 as shown in Figure 6.21. The cut-off

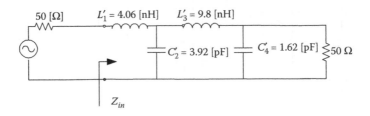

FIGURE 6.20 Final LPF with binomial response.

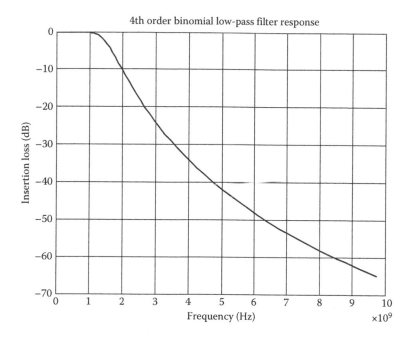

FIGURE 6.21   MATLAB results for fourth-order LPF with binomial response.

frequency with MATLAB is 1.5 GHz and it is confirmed with the frequency domain circuit simulator. The circuit that is simulated with Ansoft Designer for the fourth-order LPF and the simulation results are shown in Figures 6.22 and 6.23.

On the basis of the filter responses in Figures 6.21 through 6.23, the filter has a cut-off frequency at 1.5 GHz with 3 dB attenuation and meets the attenuation requirement at 4.5 GHz by having –38.15 dB attenuation at that level.

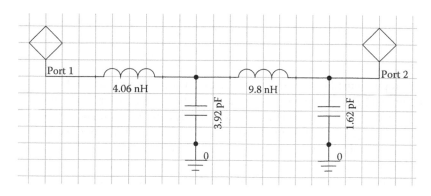

FIGURE 6.22   Simulated fourth-order LPF.

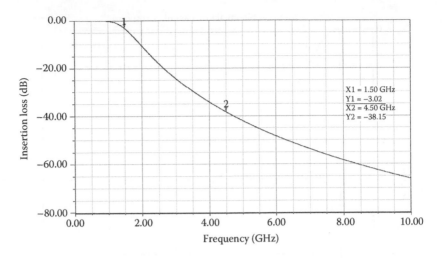

FIGURE 6.23   Simulation results for fourth-order LPF.

6.3.1.2   Chebyshev Filter Response

The Chebyshev filter response can be obtained similarly using the two-element low-pass prototype circuit given in Figure 6.16. The power loss from Equation 6.39 takes the following form when $N = 2$:

$$P_{LR} = 1 + k^2 T_2^2 \left(\frac{\omega}{\omega_c} \right) \tag{6.50}$$

where

$$T_2 \left(\frac{\omega}{\omega_c} \right) = 2 \left(\frac{\omega}{\omega_c} \right)^2 - 1 \tag{6.51}$$

Substituting Equation 6.51 into Equation 6.50 gives

$$P_{LR} = 1 + k^2 \left(4\omega^4 - 4\omega^2 + 1 \right) \tag{6.52}$$

The input impedance and S_{11} are given by Equations 6.41 and 6.42. When Equation 6.42 is substituted into Equation 6.36 with Equation 6.52, the L and C values satisfying the equation are found. Chebyshev polynomials up to the seventh order are given in Table 6.2. Polynomials given in Table 6.2 can be used to obtain design tables giving component values for various ripple values as shown in Table 6.3.

Chebyshev polynomials are defined by a three-term recursion where

$$T_0(x) = 1, \quad T_1(x) = x, \quad T_{n+1}(x) = 2xT_n(x) - T_{n-1}(x),$$
$$n = 1, 2, \dots \text{ and } x = \omega/\omega_c. \tag{6.53}$$

TABLE 6.2
Chebyshev Polynomials up to Seventh Order

Order of Polynomial, N	$T_N\left(\dfrac{\omega}{\omega_c}\right)$
1	$\dfrac{\omega}{\omega_c}$
2	$2\left(\dfrac{\omega}{\omega_c}\right)^2 - 1$
3	$4\left(\dfrac{\omega}{\omega_c}\right)^3 - 3\left(\dfrac{\omega}{\omega_c}\right)$
4	$8\left(\dfrac{\omega}{\omega_c}\right)^4 - 8\left(\dfrac{\omega}{\omega_c}\right)^2 + 1$
5	$16\left(\dfrac{\omega}{\omega_c}\right)^5 - 20\left(\dfrac{\omega}{\omega_c}\right)^3 + 5\left(\dfrac{\omega}{\omega_c}\right)$
6	$32\left(\dfrac{\omega}{\omega_c}\right)^6 - 48\left(\dfrac{\omega}{\omega_c}\right)^4 + 18\left(\dfrac{\omega}{\omega_c}\right)^2 - 1$
7	$64\left(\dfrac{\omega}{\omega_c}\right)^7 - 112\left(\dfrac{\omega}{\omega_c}\right)^5 + 58\left(\dfrac{\omega}{\omega_c}\right)^3 - 7\left(\dfrac{\omega}{\omega_c}\right)$

The attenuation curves for the Chebyshev LPF response are obtained from Equations 6.39 and 6.44 as

$$\text{Attenuation (dB)} = 10\log\left(1 + \varepsilon^2 T_N^2\left(\frac{\omega}{\omega_c}\right)'\right) \qquad (6.54)$$

where

$$\left(\frac{\omega}{\omega_c}\right)' = \left(\frac{\omega}{\omega_c}\right)\cosh(B) \qquad (6.55)$$

$$B = \frac{1}{N}\cosh^{-1}\left(\frac{1}{\varepsilon}\right) \qquad (6.56)$$

$$\varepsilon = \sqrt{10^{\frac{ripple\,(\text{dB})}{10}} - 1} \qquad (6.57)$$

TABLE 6.3

Component Values for Chebyshev Low-Pass Filter Response with $g_0 = 1$, $\omega_c = 1$, and $N = 1$–7

N	g_1	g_2	g_3	g_4	g_5	g_6	g_7	g_8
			Ripple = 0.01 dB					
1	0.096	1						
2	0.4488	0.4077	1.1007					
3	0.6291	0.9702	0.6291	1				
4	0.7128	1.2003	1.3212	0.6476	1.1007			
5	0.7563	1.3049	1.5773	1.3049	0.7563	1		
6	0.7813	1.36	1.6896	1.535	1.497	0.7098	1.1007	
7	0.7969	1.3924	1.7481	1.6331	1.7481	1.3924	0.7969	1
			Ripple = 0.1 dB					
1	0.3052	1						
2	0.843	0.622	1.3554					
3	1.0315	1.1474	1.0315	1				
4	1.1088	1.3061	1.7703	0.818	1.3554			
5	1.1468	1.3712	1.975	1.3712	1.1468	1		
6	1.1681	1.4039	2.0562	1.517	1.9029	0.8618	1.3554	
7	1.1811	1.4228	2.0966	1.5733	2.0966	1.4228	1.1811	1
			Ripple = 0.5 dB					
1	0.6986	1						
2	1.4029	0.7071	1.9841					
3	1.5963	1.0967	1.5963	1				
4	1.6703	1.1926	2.3661	0.8419	1.9841			
5	1.7058	1.2296	2.5408	1.2296	1.7058	1		
6	1.7254	1.2479	2.6064	1.3137	2.4758	0.8696	1.9841	
7	1.7372	1.2583	2.6381	1.3444	2.6381	1.2583	1.77372	
			Ripple = 1 dB					
1	1.0177	1						
2	1.8219	0.685	2.6599					
3	2.0236	0.9941	2.0236	1				
4	2.0991	1.0644	2.8311	0.7892	2.6599			
5	2.1349	1.0911	3.0009	1.0911	2.1349	1		
6	2.1546	1.1041	3.0634	1.1518	2.9367	0.8101	2.6599	
7	2.1664	1.1116	3.0934	1.1736	3.0934	1.1116	2.1664	1
			Ripple = 3 dB					
1	1.9953	1						
2	3.1013	0.5339	5.8095					
3	3.3487	0.7117	3.3487	1				
4	3.4389	0.7483	4.3471	0.592	5.8095			
5	3.4817	0.7618	4.5381	0.7618	3.4817	1		
6	3.5045	0.7685	4.6061	0.7929	4.4641	0.6033	5.8095	
7	3.5182	0.7723	4.6386	0.8039	4.6386	0.7723	3.5182	1

The attenuation curves are obtained and given for several ripple values using MATLAB as shown in Figures 6.24 through 6.28.

Example 2: *F/2* Low-Pass Filter Design for RF Power Amplifiers

Design an *F/2* filter for RF power amplifier that is operating at 13.56 MHz. The filter should have no impact during the normal operation of the amplifier. It should have at least 20 dB attenuation at *F/2* frequency. The passband ripple should not exceed 0.1 dB. It is given that the amplifier is presenting 30 Ω impedance to the filter on its load line.

SOLUTION

In RF power amplifier applications, signals having a frequency of *F/2* may become an important problem that affects the signal purity and the amount of power delivered to the load. This problem can be resolved by eliminating signals using LPFs commonly called as *F/2* filter. *F/2* filter is connected offline to the load line of the amplifier and presents high impedance at the center frequency but matched impedance at *F/2*. The analysis begins with identifying *F/2* frequency as

$$\frac{F}{2} = 6.78 \, [\text{MHz}]$$

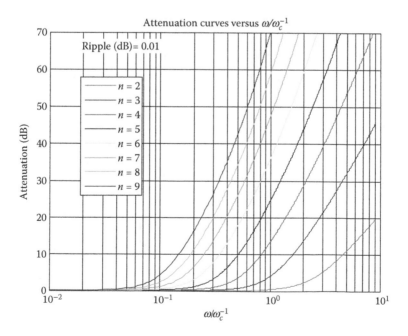

FIGURE 6.24 Attenuation curves for Chebyshev filter response for 0.01 dB ripple.

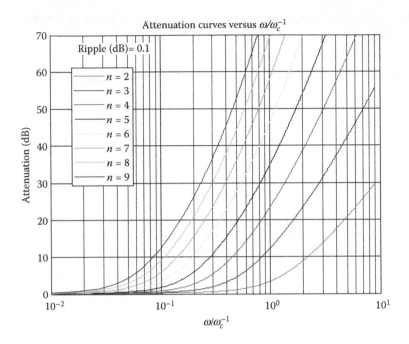

FIGURE 6.25 Attenuation curves for Chebyshev filter response for 0.1 dB ripple.

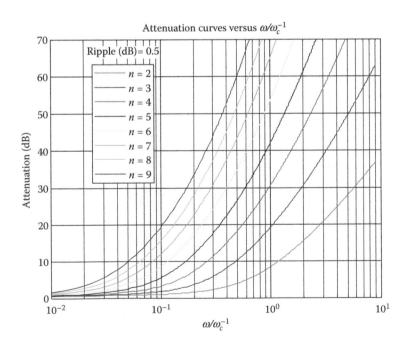

FIGURE 6.26 Attenuation curves for Chebyshev filter response for 0.5 dB ripple.

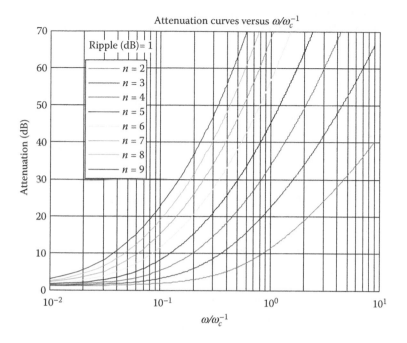

FIGURE 6.27 Attenuation curves for Chebyshev filter response for 1 dB ripple.

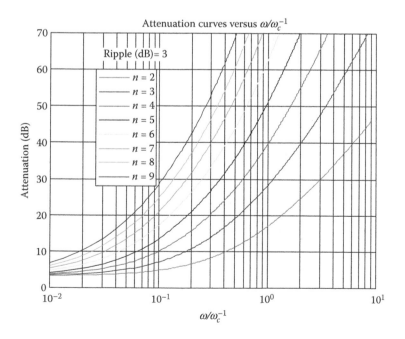

FIGURE 6.28 Attenuation curves for Chebyshev filter response for 3 dB ripple.

The cut-off frequency of the filter is selected to be 25–35% higher than $F/2$ as a rule of thumb. The attenuation at the cut-off frequency is expected to be 3 dB as shown below:

$$\text{Attenuation} = 3 \text{ dB} \quad \text{at} \quad f_c = 9 \text{ [MHz]}$$

Now, we can apply the steps that we used earlier to design the filter. Since the ripple requirement in the passband is mentioned, the Chebyshev filter topology is used to design and implement the filter.

Step 1—Use Figure 6.25 to determine the required number of elements to obtain a minimum 20 dB attenuation at $f = 13.56$ MHz.

$$\left| \frac{\omega}{\omega_c} \right| - 1 = \frac{13.56}{9} - 1 = 0.5 \rightarrow N = 5$$

Step 2—Use Table 6.3 to determine the normalized LPF component values as

		Ripple = 0.1 dB			
N	g_1	g_2	g_3	g_4	g_5	g_6
5	1.1468	1.3712	1.975	1.3712	1.1468	1

The normalized component values for the filter are obtained as shown in Figure 6.29 from the table as

$$L_1 = L_5 = 1.1468, \quad L_3 = 1.975, \quad C_2 = C_4 = 1.3712$$

Step 3—Apply impedance and frequency scaling

$$R_s' = R_o = 30 \text{ [}\Omega\text{]}$$

$$R_L' = R_o R_L = 30(1) = 30 \text{ [}\Omega\text{]}$$

$$L_1' = L_5' = R_o \frac{L_n}{\omega_c} = 30 \frac{1.1468}{(2\pi \times 8 \times 10^6)} = 684.44 \text{ [nH]}$$

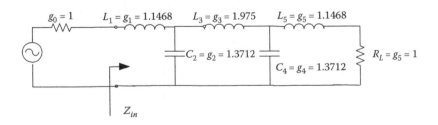

FIGURE 6.29 Fifth-order normalized LPF for Chebyshev response.

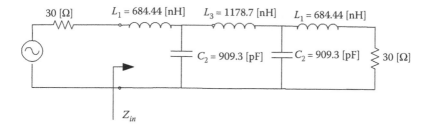

FIGURE 6.30 Final LPF with Chebyshev response.

$$C_2' = C_4' = \frac{C_n}{R_o \omega_c} = \frac{1.3712}{30(2\pi \times 8 \times 10^6)} = 909.3 \text{ [pF]}$$

$$L_3' = R_o \frac{L_n}{\omega_c} = 30 \frac{1.975}{(2\pi \times 8 \times 10^6)} = 1178.7 \text{ [nH]}$$

The final LPF circuit having the Chebyshev filter response is shown in Figure 6.30. The final circuit shown in Figure 6.30 is analyzed using network parameters and the insertion loss is obtained with $ABCD$ parameters for the cascaded components as previously discussed.

$$\begin{bmatrix} A & B \\ C & D \end{bmatrix} = \begin{bmatrix} 1 & Z_S \\ 0 & 1 \end{bmatrix} \begin{bmatrix} 1 & j\omega L_1' \\ 0 & 1 \end{bmatrix} \begin{bmatrix} 1 & 0 \\ j\omega C_2' & 1 \end{bmatrix} \begin{bmatrix} 1 & j\omega L_3' \\ 0 & 1 \end{bmatrix}$$
$$\times \begin{bmatrix} 1 & 0 \\ j\omega C_4' & 1 \end{bmatrix} \begin{bmatrix} 1 & j\omega L_5' \\ 0 & 1 \end{bmatrix} \begin{bmatrix} 1 & 0 \\ \dfrac{1}{Z_L} & 1 \end{bmatrix} \qquad (6.58)$$

The insertion loss in the passband and stopband are obtained using MATLAB from (6.58) and are shown in Figures 6.31 and 6.32.

The passband ripple is less than 0.1 dB and the cut-off frequency is around 9 MHz as shown in Figure 6.31. In addition, we have more than 25 dB attenuation at 13.56 MHz as illustrated in Figure 6.32. The circuit is simulated with Ansoft Designer for accuracy using the circuit shown in Figure 6.33. The passband ripple, the attenuation at cut-off frequency, and the operational frequency are given in Figure 6.34 and are in agreement with the MATLAB results obtained.

The input impedance for the filter designed is given in Figure 6.35. On the basis of the results on the Smith chart, the filter input impedance is $29.58 - j8.48\ \Omega$ at $F/2$ and $0.06 + j43.02\ \Omega$ at F. Hence, the filter presents very closely matched load to the amplifier at $F/2$ and terminates the $F/2$ frequency content and presents a very high inductance and acts like an open load at the operational frequency and does not have any impact on the amplifier performance.

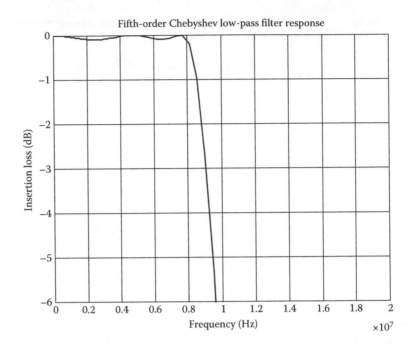

FIGURE 6.31   Passband ripple response for fifth-order LPF with Chebyshev filter response.

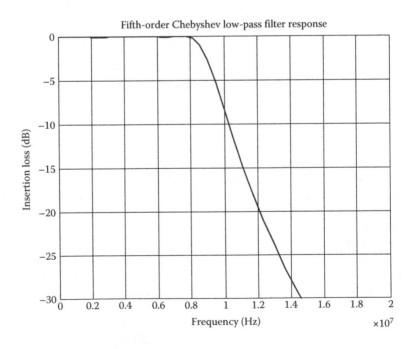

FIGURE 6.32   Attenuation response for fifth-order LPF with Chebyshev filter response.

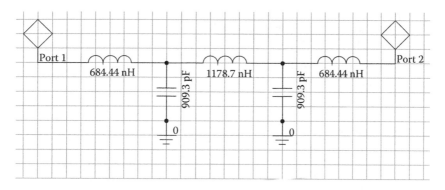

FIGURE 6.33   Simulated fifth-order LPF.

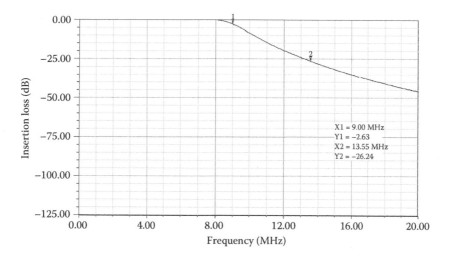

FIGURE 6.34   Simulation results for fifth-order LPF.

6.3.2 HIGH-PASS FILTERS

HPFs are designed from LPF prototypes using the frequency transformation given by

$$-\frac{\omega_c}{\omega} \to \omega \tag{6.59}$$

This transformation converts LPF to HPF with the following frequency and impedance scaling relations for L and C

$$L'_n = \frac{R_o}{\omega_c C_n} \tag{6.60}$$

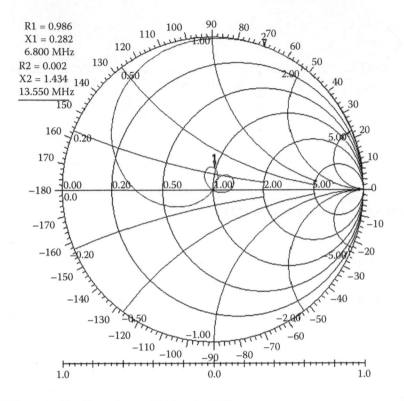

R1 = 0.986
X1 = 0.282
6.800 MHz
R2 = 0.002
X2 = 1.434
13.550 MHz

FIGURE 6.35 Input impedance of fifth-order LPF.

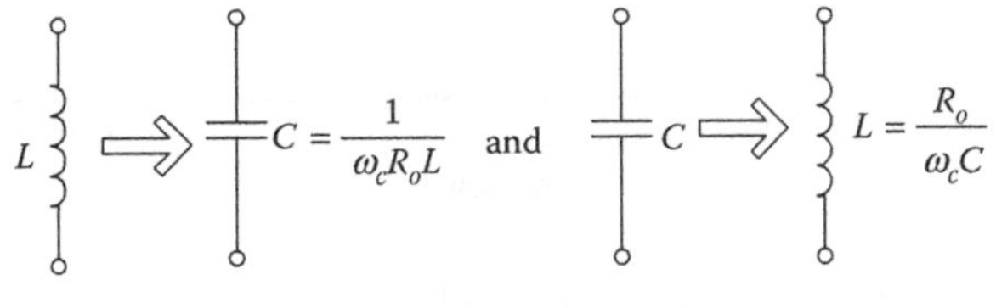

$$L \Rightarrow C = \frac{1}{\omega_c R_o L} \quad \text{and} \quad C \Rightarrow L = \frac{R_o}{\omega_c C}$$

FIGURE 6.36 LPF component to HPF component transformation.

$$C'_n = \frac{1}{\omega_c R_o L_n} \tag{6.61}$$

The design begins with the low-pass prototype by finding L_n and C_n, and then applies (6.60) and (6.61). The transformation of the components from LPF to HPF is illustrated in Figure 6.36.

Example 3: High-Pass Filter Design

Design an HPF with a 3 dB equal ripple response and a cut-off frequency of 1 GHz. The source and load impedances are given to be 50 Ω and the attenuation at 0.6 GHz is required to be a minimum 40 dB.

SOLUTION

Step 1—Use Figure 6.28 to determine the required number of elements to obtain a minimum 40 dB attenuation at $f = 0.6$ MHz. Apply Equation 6.59 and obtain

$$\left| \left(\frac{-\frac{\omega_c}{\omega}}{\omega_c} \right) \right| - 1 = \frac{1}{0.6} - 1 = 0.667 \rightarrow N = 5$$

Step 2—Use Table 6.3 to determine the normalized LPF component values as

			Ripple = 3 dB			
N	g_1	g_2	g_3	g_4	g_5	g_6
5	3.4817	0.7618	4.5381	0.7618	3.4817	1

We would like to begin with the shunt capacitor as the first element in the LPF prototype. However, the LPF has to be transformed to the HPF as shown in Figure 6.37.

Now, the HPF component values can be calculated using Equations 6.60 and 6.61 as

$$L_1' = L_5' = \frac{50}{(2\pi \times 1 \times 10^9)(3.4817)} = 2.28 \text{ [nH]}$$

$$C_2' = C_4' = \frac{1}{(2\pi \times 1 \times 10^9)(50)(0.7618)} = 4.18 \text{ [pF]}$$

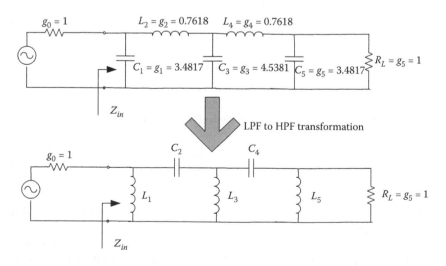

FIGURE 6.37 LPF prototype circuit to HPF transformation.

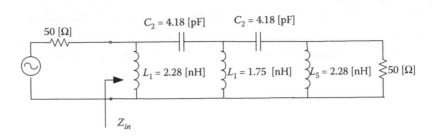

FIGURE 6.38   Final HPF filter.

$$L_3' = \frac{50}{(2\pi \times 1 \times 10^9)(4.5381)} = 1.75 \ [nH]$$

The final circuit is shown in Figure 6.38 and its response is obtained using MATLAB as it was done earlier using network parameters as

$$\begin{bmatrix} A & B \\ C & D \end{bmatrix} = \begin{bmatrix} 1 & Z_S \\ 0 & 1 \end{bmatrix} \begin{bmatrix} 1 & 0 \\ 1/j\omega L_1' & 1 \end{bmatrix} \begin{bmatrix} 1 & 1/j\omega C_2' \\ 0 & 1 \end{bmatrix} \begin{bmatrix} 1 & 0 \\ 1/j\omega L_3' & 1 \end{bmatrix}$$

$$\times \begin{bmatrix} 1 & 1/j\omega C_4' \\ 0 & 1 \end{bmatrix} \begin{bmatrix} 1 & 0 \\ 1/j\omega L_5' & 1 \end{bmatrix} \begin{bmatrix} 1 & 0 \\ \dfrac{1}{Z_L} & 1 \end{bmatrix} \tag{6.62}$$

The attenuation response obtained by MATLAB is shown in Figure 6.39. The attenuations at cut-off frequency 1 and 0.6 GHz are around 3 and 40 dB as expected. The circuit is simulated with Ansoft Designer for accuracy using the circuit shown in Figure 6.40. The passband ripple, the attenuation at cut-off frequency, and the operational frequency are given in Figure 6.41. It has been shown that 3 dB attenuation at 1 GHz and 42.14 dB attenuation at 0.6 GHz are obtained with the HPF designed.

6.3.3   BANDPASS FILTERS

BPFs are designed from LPF prototypes using the frequency transformation given by

$$\frac{\omega_o}{\omega_{c2} - \omega_{c1}} \left(\frac{\omega}{\omega_o} - \frac{\omega_o}{\omega} \right) \rightarrow \omega \tag{6.63}$$

The term $(\omega_{c2} - \omega_{c1})/\omega_o$ is called the fractional bandwidth and ω_o is called the resonant or center frequency and is defined by Equation 6.18. ω_{c2} and ω_{c1} are the upper and lower cut-off frequencies and are defined by Equation 6.17. The transformation given by Equation 6.63 maps the series component of the LPF prototype circuit to series LC circuit and the shunt component of the LPF prototype circuit

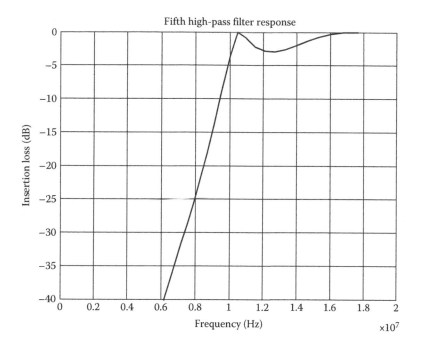

FIGURE 6.39 Attenuation response for fifth-order HPF.

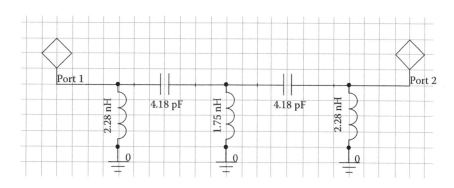

FIGURE 6.40 Simulated fifth-order HPF.

to shunt LC circuit in the BPF. The component values of the series LC circuit are calculated as

$$L'_n = \frac{R_o L_n}{(\omega_o/(\omega_{c2} - \omega_{c1}))^{-1} \omega_o} \tag{6.64}$$

$$C'_n = \frac{(\omega_o/(\omega_{c2} - \omega_{c1}))^{-1}}{\omega_o R_o L_n} \tag{6.65}$$

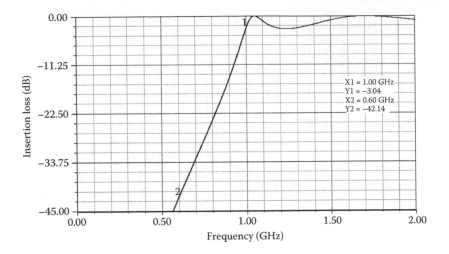

FIGURE 6.41   Simulation results for fifth-order LPF.

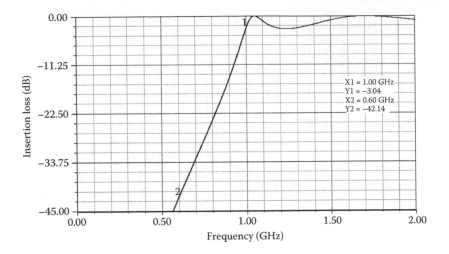

FIGURE 6.42   LPF component to BPF component transformation.

The component values of the shunt LC circuit are calculated as

$$L_n' = \frac{(\omega_o/(\omega_{c2} - \omega_{c1}))^{-1} R_o}{\omega_o C_n} \tag{6.66}$$

$$C_n' = \frac{C_n}{(\omega_o/\omega_{c2} - \omega_{c1})^{-1} R_o \omega_o} \tag{6.67}$$

The transformation of the components from LPF to BPF is illustrated in Figure 6.42.

Example 4: Bandpass Filter Design

Design a BPF with 5% fractional bandwidth and center frequency of 2 GHz. The filter is desired to have maximally flat response in the passband and have four sections. The source and load impedances are given to be 50 Ω.

SOLUTION

The filter specifications mention that the BPF filter has four sections and maximally flat bandpass response. As a result, we will be using binomial LPF filter prototype circuit to design the BPF circuit to meet specifications. The fractional bandwidth is given to be $(\omega_o / (\omega_{c2} - \omega_{c1}))^{-1} = 0.05$.

Step 1—The required number of sections is defined as four. This requires the LPF prototype to have four components. So, $N = 4$.

Step 2—Use Table 6.1 to determine the normalized LPF component

N	g_1	g_2	g_3	g_4	g_5
4	0.7654	1.8478	1.8478	0.7654	1.0000

We would like to begin with the series inductor as the first element in the LPF prototype. However, LPF has to be transformed to BPF as shown in Figure 6.43.

The BPF component values can be calculated using Equations 6.64 through 6.67 as

$$L_1' = \frac{50(0.7654)}{0.05(2\pi \times 2 \times 10^9)} = 60.908 \text{ [nH]}$$

$$C_1' = \frac{0.05}{(2\pi \times 2 \times 10^9)50(0.7654)} = 0.10396 \text{ [pF]}$$

$$L_2' = \frac{(0.05)50}{(2\pi \times 2 \times 10^9)(1.8478)} = 0.107 \text{ [nH]}$$

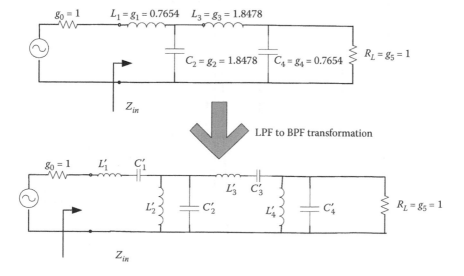

FIGURE 6.43 LPF prototype circuit to BPF transformation.

$$C_2' = \frac{1.8478}{(0.05)(50)(2\pi \times 2 \times 10^9)} = 58.817 \text{ [pF]}$$

$$L_3' = \frac{50(1.8478)}{0.05(2\pi \times 2 \times 10^9)} = 147.04 \text{ [nH]}$$

$$C_3' = \frac{0.05}{(2\pi \times 2 \times 10^9)50(1.8478)} = 0.043 \text{ [pF]}$$

$$L_4' = \frac{(0.05)50}{(2\pi \times 2 \times 10^9)(0.7654)} = 0.258 \text{ [nH]}$$

$$C_4' = \frac{0.7654}{(0.05)(50)(2\pi \times 2 \times 10^9)} = 24.363 \text{ [pF]}$$

The final circuit is analyzed using MATLAB as it was done earlier using network parameters and the response is obtained as shown in Figure 6.44. BPF is simulated with Ansoft Designer using the circuit in Figure 6.45. Ansoft Designer attenuation response confirming the results of MATLAB for BPF is given in Figure 6.46.

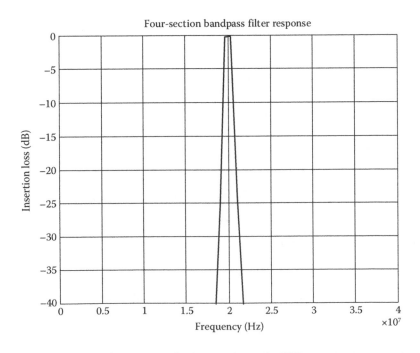

FIGURE 6.44 Attenuation response for four-section-order BPF.

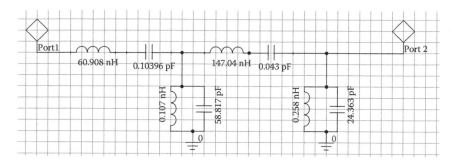

FIGURE 6.45 Simulated four-section BPF.

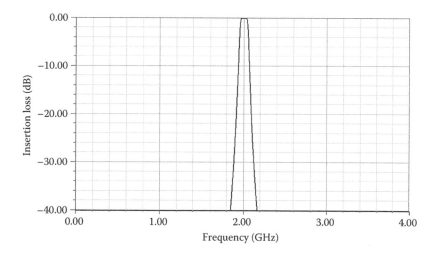

FIGURE 6.46 Simulation results for four-section BPF.

6.3.4 BANDSTOP FILTERS

BSFs are designed from LPF prototypes using the frequency transformation given by

$$\frac{\omega_{c2} - \omega_{c1}}{\omega_o}\left(\frac{\omega}{\omega_o} - \frac{\omega_o}{\omega}\right)^{-1} \to \omega \qquad (6.68)$$

This transformation maps the series component of the LPF prototype circuit to the shunt LC circuit and the shunt component of the LPF prototype circuit to the series LC circuit in the BSF. The component values of the shunt LC circuit are calculated as

$$L'_n = \frac{(\omega_o/(\omega_{c2} - \omega_{c1}))^{-1} L_n R_0}{\omega_0} \qquad (6.69)$$

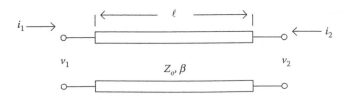

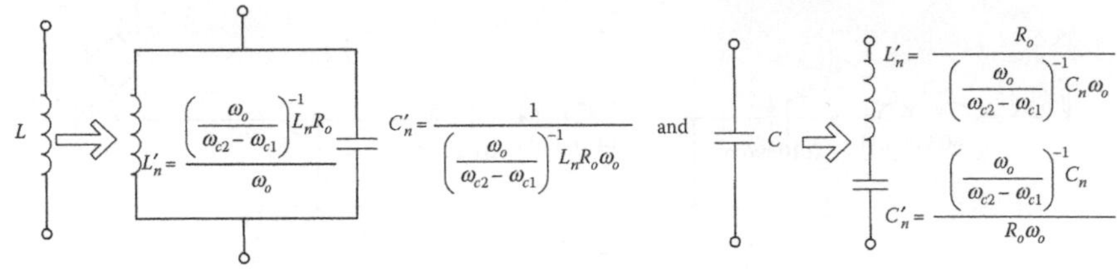

FIGURE 6.47 LPF component to BSF component transformation.

$$C'_n = \frac{1}{\left(\omega_o/(\omega_{c2} - \omega_{c1})\right)^{-1} L_n R_o \omega_o} \tag{6.70}$$

The component values of the series LC circuit are calculated as

$$L'_n = \frac{R_o}{\left(\omega_o/(\omega_{c2} - \omega_{c1})\right)^{-1} C_n \omega_o} \tag{6.71}$$

$$C'_n = \frac{\left(\omega_o/(\omega_{c2} - \omega_{c1})\right)^{-1} C_n}{R_o \omega_o} \tag{6.72}$$

The transformation of the components from LPF to BSF is illustrated in Figure 6.47.

6.4 STEPPED IMPEDANCE LOW-PASS FILTERS

A stepped impedance filter is made up of high- and low-impedance sections of the transmission line shown in Figure 6.48. Using the transmission line theory, the high-impedance sections and the low-impedance sections are implemented to realize LPFs.

The two-port Z-parameter matrix for a transmission line in Figure 6.48 is

$$Z = \begin{bmatrix} -jZ_o \cot(\beta\ell) & -jZ_o \csc(\beta\ell) \\ -jZ_o \csc(\beta\ell) & -jZ_o \cot(\beta\ell) \end{bmatrix} \tag{6.73}$$

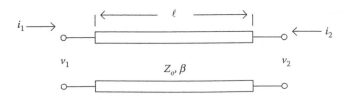

FIGURE 6.48 Transmission line model.

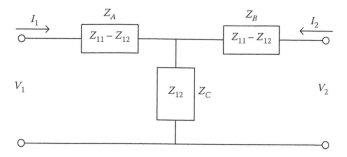

FIGURE 6.49 T-network equivalent circuit.

where

$$Z_{11} = Z_{22} = -jZ_o \cot(\beta\ell) \quad \text{and} \quad Z_{12} = Z_{21} - jZ_o \csc(\beta\ell) \tag{6.74}$$

An equivalent T-connected network can be used to represent the two-port transmission line network in Figure 6.48. The equivalent T-connected network representing the transmission line is shown in Figure 6.49.

The components of the T-network are defined as

$$Z_A = Z_B = Z_{11} - Z_{12} = jZ_o \tan\left(\frac{\beta\ell}{2}\right) \tag{6.75}$$

$$Z_C = Z_{12} = -jZ_o \csc(\beta\ell) \tag{6.76}$$

When the electrical length, $\beta\ell$, is small, then the following approximation can be done:

$$\sin(\beta\ell) \approx \beta\ell, \quad \cos(\beta\ell) \approx 1, \quad \text{and} \quad \tan(\beta\ell) \approx \beta\ell \tag{6.77}$$

The approximations given by Equation 6.77 lead to the following element values for the T-network as shown in Figure 6.50:

$$Z_A = Z_B = Z_{11} - Z_{12} \approx jZ_o\left(\frac{\beta\ell}{2}\right) \tag{6.78}$$

$$Z_C = Z_{12} \approx \frac{Z_o}{j\beta\ell} \tag{6.79}$$

Consider the case when the characteristic impedance Z_o is very high. We denote this impedance as Z_{High}. For the shunt component, since $\beta\ell$ is very small, the impedance will be very large. In fact, it can be considered as an open circuit. This results

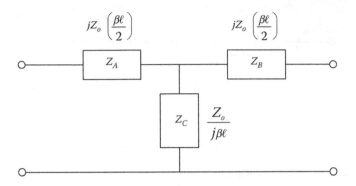

FIGURE 6.50   T-network representation with transmission lines.

in an approximate circuit impedance of the series component, $jZ_{High}\, \beta\ell$, as

$$\frac{Z_{High}}{j\beta\ell} \to \infty \quad \text{when } \beta\ell \ll 1 \quad \text{and} \quad Z_{High} \gg Z_o \qquad (6.80)$$

High-impedance condition transforms the T-network to equivalent series-connected L-network as shown in Figure 6.51. Now, consider the case when the characteristic impedance is low (Z_{Low}). This time, the series components have a very low impedance and can be considered short. The resulting approximate circuit impedance is that of the shunt component alone or $Z_{Low}/j\beta\ell$ as

$$jZ_{Low}\left(\frac{\beta\ell}{2}\right) \to 0 \quad \text{when } \beta\ell \ll 1 \quad \text{and} \quad Z_{Low} \ll Z_0 \qquad (6.81)$$

As a result, the low-impedance condition transforms the T-network to equivalent shunt-connected C-network as shown in Figure 6.52. The physical length of the component values for series- and shunt-connected elements are found from Equations 6.80 and 6.81. The length for the inductive element can be obtained from

$$X_L = j\omega L = jZ_{High}\, \beta\ell \qquad (6.82)$$

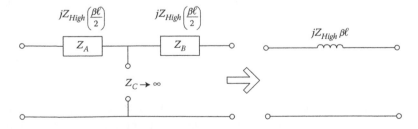

FIGURE 6.51   High-impedance transformation of T-network.

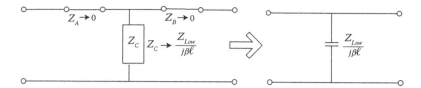

FIGURE 6.52 Low-impedance transformation of T-network.

So

$$\ell_{High} = \frac{\omega L}{Z_{High}\beta}$$ (6.83)

The length for the capacitive elements are found from

$$X_C = \frac{1}{j\omega C} = \frac{Z_{Low}}{j\beta\ell}$$ (6.84)

and the length is

$$\ell_{Low} = \frac{Z_{Low}\omega C}{\beta}.$$ (6.85)

L and C values are the values obtained using LPF prototype circuit based on the filter specifications. In Equations 6.82 and 6.85, the phase constant is defined as

$$\beta = \frac{\omega}{v_p}$$ (6.86)

where v_p is a phase velocity as defined by

$$v_p = \frac{c}{\sqrt{\varepsilon_e}}$$ (6.87)

ε_e is the effective permittivity cons of the microstrip line. The high- and low-impedance values are desired to be

$$Z_{Low} < Z_o < Z_{High}$$ (6.88)

The selection of Z_{Low} and Z_{High} values carries importance for the response of the filter. The ratio of Z_{High} to Z_{Low} should be kept as large as possible to obtain more accurate results. We can define the approximate limits for Z_{Low} and Z_{High} based on the assumption that the electrical length is small if

$$\beta\ell < \frac{\pi}{4}$$ (6.89)

Then, the impedance limit for Z_{Low} is

$$Z_{Low} < \frac{\pi}{4\omega_c C} \tag{6.90}$$

and the impedance limit for Z_{High} is

$$Z_{High} > \frac{4\omega_c L}{\pi} \tag{6.91}$$

Once Z_{Low} and Z_{High} are defined, the width of each line can be obtained using microstrip line equation defined by

$$\frac{W}{d} = \begin{cases} \dfrac{8e^A}{e^{2A} - 2} & \text{for } W/d < 2 \\[2ex] \dfrac{2}{\pi}\left[B - 1 - \ln(2B-1) + \dfrac{\varepsilon_r - 1}{2\varepsilon_r}\left\{ \ln(B-1) + 0.39 - \dfrac{0.61}{\varepsilon_r}\right\}\right] & \text{for } W/d > 2 \end{cases} \tag{6.92}$$

where

$$A = \frac{Z_o}{60}\sqrt{\frac{\varepsilon_r + 1}{2}} + \frac{\varepsilon_r - 1}{\varepsilon_r + 1}\left(0.23 + \frac{0.11}{\varepsilon_r}\right) \tag{6.93}$$

$$B = \frac{377\pi}{2Z_o\sqrt{\varepsilon_r}} \tag{6.94}$$

Design Example: Design, Simulation, and Implementation of Stepped Impedance Filter

Design, simulate, and implement a stepped impedance LPF with a cut-off frequency at 2 GHz. The filter is desired to provide a minimum 30 dB attenuation at 3 GHz. The source and load impedances of the filter are given to be 50 Ω. The ripple is defined to be not more than 0.5 dB in the passband. In addition, it is required to use FR4 as substrate with a dielectric constant of 3.7 and a dielectric thickness of 60 mil.

SOLUTION

It is mentioned in the filter specifications that there is a ripple requirement in the passband. This is an indication of the Chebyshev-type filter.

Step 1—Use Figure 6.26 to determine the required number of elements to obtain a minimum 30 dB attenuation at $f = 3$ GHz.

$$\left|\frac{\omega}{\omega_c}\right| - 1 = \frac{3}{2} - 1 = 0.5 \rightarrow N = 7$$

Step 2—Use Table 6.3 to determine the normalized LPF component values with 0.5 dB ripple as

				Ripple = 0.5 dB				
N	g_1	g_2	g_3	g_4	g_5	g_6	g_7	g_8
7	1.7372	1.2583	2.6381	1.3444	2.6381	1.2583	1.77372	1

We begin the prototype LPF circuit with shunt C as shown in Figure 6.53. The normalized component values for the filter are obtained from the table as

$$C_1 = C_7 = 1.7372, \quad C_3 = C_5 = 2.6381, \quad L_2 = L_6 = 1.2583, \quad L_4 = 1.3444$$

Step 3—Apply the scaling impedance and frequency

$$R_s' = R_o = 50 \; [\Omega]$$

$$R_L' = R_o R_L = 50(1) = 50 \; [\Omega]$$

$$C_1' = C_7' = \frac{1.7372}{50(2\pi \times 2 \times 10^9)} = 2.764 \; [pF]$$

$$L_2' = L_6' = 50 \frac{1.2583}{(2\pi \times 2 \times 10^9)} = 5 \; [nH]$$

$$C_3' = C_5' = \frac{2.6381}{50(2\pi \times 2 \times 10^9)} = 4.197 \; [pF]$$

$$L_4' = 50 \frac{1.3444}{(2\pi \times 2 \times 10^9)} = 5.342 \; [nH]$$

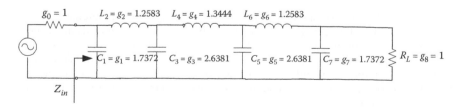

FIGURE 6.53 Seventh-order normalized LPF for Chebyshev response.

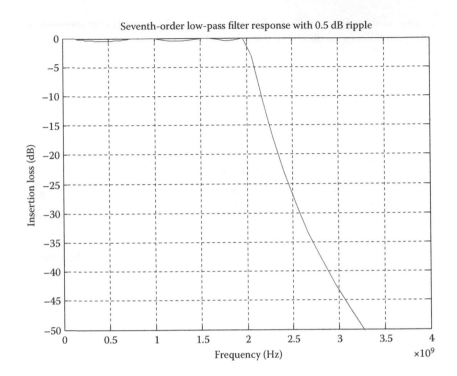

FIGURE 6.54   Attenuation response of seventh-order Chebyshev LPF.

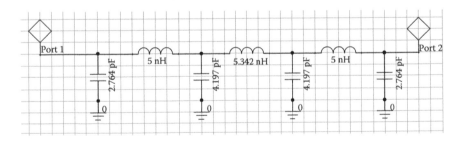

FIGURE 6.55   Simulated seventh-order Chebyshev LPF.

The MATLAB response of this circuit is shown in Figure 6.54. This circuit is simulated with Ansoft Designer for its frequency response and the simulated circuit and its response versus frequency are illustrated in Figures 6.55 and 6.56, respectively. The simulation results confirm the MATLAB results, and attenuation requirements at the cut-off frequency and 3 GHz are also met. At this point, we can go ahead and move to step 4 to transform our filter to step impedance filter using the design method described.

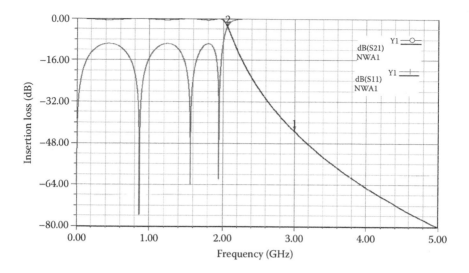

FIGURE 6.56   Simulation result for seventh-order Chebyshev LPF with 0.5 dB ripple.

Step 4—Choose Z_{Low} and Z_{High}.
The low impedance limit is found to be

$$Z_{Low} < \frac{\pi}{4\omega_c C} \rightarrow Z_{Low} < 14.88 \, [\Omega] \tag{6.95}$$

and the high impedance limit is found to be

$$Z_{High} > \frac{4\omega_c L}{\pi} \rightarrow Z_{High} > 85.47 \, [\Omega] \tag{6.96}$$

On the basis of the defined limits obtained by Equations 6.95 and 6.96, we set our impedances as

$$Z_{Low} = 14 \, [\Omega] \quad \text{and} \quad Z_{High} = 120 \, [\Omega] \tag{6.97}$$

Step 5—Determine the physical dimensions of the transmission lines.

MATLAB program has been developed to obtain the physical dimensions of the transmission lines in the stepped impedance filter. The physical dimensions obtained by MATLAB are

$l_1 = 9.5$ mm, $l_2 = 5.2$ mm, $l_3 = 14.4$ mm, $l_4 = 5.6$ mm,
$l_5 = 14.4$ mm, $l_6 = 5.2$ mm, $l_7 = 9.5$ mm
$w_1 = 17.6$ mm, $w_2 = 0.5$ mm, $w_3 = 17.6$ mm, $w_4 = 0.5$ mm,
$w_5 = 17.6$ mm, $w_6 = 0.5$ mm, $w_7 = 17.6$ mm

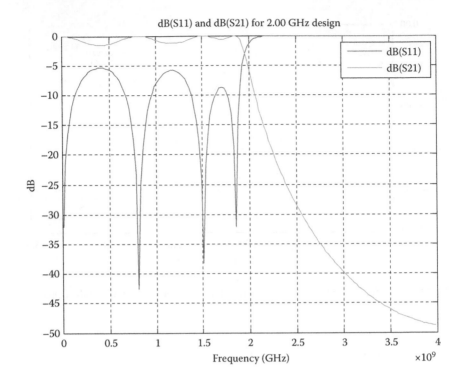

FIGURE 6.57   Attenuation profile for the step impedance filter.

The attenuation profile obtained by MATLAB is shown in Figure 6.57.

Comparison of the attenuation profiles obtained by ideal Chebyshev LPF and step impedance filter response by MATLAB are in agreement. It is important to note that a slight difference between the two profiles is due to the approximation applied in the stepped impedance filter design.

Step 6—Optimize and obtain the final design with the electromagnetic simulator.

We can now use the electromagnetic simulator to optimize this design before implementation. Sonnet planar electromagnetic simulator is used to optimize the final physical dimensions of the filter. The final dimensions of the filter are given in Table 6.4. The simulated structure is illustrated in Figure 6.58.

The simulation results of the Sonnet are shown in Figure 6.59. It is shown that the desired attenuation profile is obtained with the optimized physical dimensions given in Table 6.4.

Step 7—Implement the filter.

Since the final design now meets the required attenuation profile, we can implement the design. The implemented structure is shown in Figure 6.60.

The measured values showing the insertion loss of the filter are given in Figure 6.61. It is shown from Figure 6.61 that the attenuation at 2 GHz is 2.8 dB and more than 30 dB at 3 GHz.

TABLE 6.4
Final Dimensions of Step Impedance LPF

Element	Width (mm)	Length (mm)
R_s	3.25	10
C_1	11.5	7.8
L_2	0.48	6.0
C_3	11.5	11.8
L_4	0.48	6.4
C_5	11.5	11.8
L_6	0.48	6.0
C_7	11.5	7.8
R_L	3.25	10

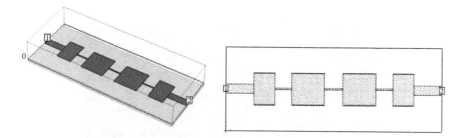

FIGURE 6.58   Simulated step impedance LPF structure.

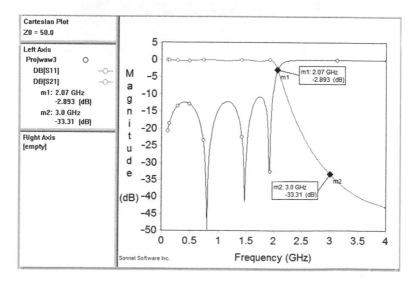

FIGURE 6.59   Simulation results of the step impedance LPF structure with Sonnet.

FIGURE 6.60   Step impedance filter is implemented.

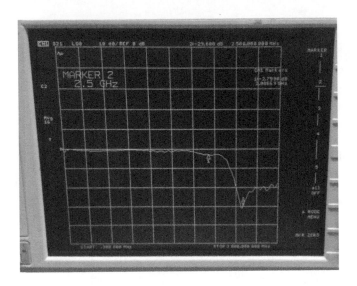

FIGURE 6.61   Measured results for the step impedance filter.

6.5   STEPPED IMPEDANCE RESONATOR BANDPASS FILTERS

The conventional parallel coupled BPFs suffer drastically from the spurious harmonics. The stepped impedance resonator filters (SIRs) can be used to realize high-performance BPFs by suppressing the spurious harmonics to overcome this problem. One of the key features of an SIR is that its resonant frequencies can be tuned by adjusting the impedance ratios of the high-Z and low-Z sections. The symmetrical trisection SIR used in the BPF design is shown in Figure 6.62.

In the symmetrical SIR structure, each section is desired to have the same electrical length. Then, it can be shown that the resonance occurs when it is equal to

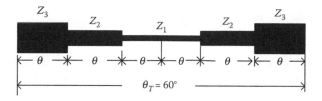

FIGURE 6.62 Three-section SIR.

$$\theta = \tan^{-1}\left(\sqrt{\frac{K_1 K_2}{K_1 + K_2 + 1}}\right) \tag{6.98}$$

where

$$K_1 = \frac{-(\cos\alpha)(\cos\beta) + \sqrt{(\cos\alpha)^2 (\cos\beta)^2 + 4(\sin b)^2 (\cos(ab))^2}}{2(\cos(ab))^2} \tag{6.99}$$

$$K_2 = \frac{1 + K_1}{\tan^2(ab) - K_1} \tag{6.100}$$

The design parameters in Equations 6.99 and 6.100 are found from

$$a = \frac{f_{s1}}{f_o} \tag{6.101a}$$

$$b = \frac{\pi}{2}\frac{f_o}{f_{s2}} \tag{6.101b}$$

$$\alpha = \frac{\pi}{2}\frac{f_{s1} + f_o}{f_{s2}} \tag{6.102a}$$

$$\beta = \frac{\pi}{2}\frac{f_{s1} - f_o}{f_{s2}} \tag{6.102b}$$

The terminating impedance of the SIR at the input and output is desired to be $Z_3 = 50$ [Ω]. Once the operating frequencies, f_o, f_{s1}, and f_{s2}, of the BPF and the terminating impedance, Z_3, of SIR are identified, the line impedances, Z_1 and Z_2, are found from

$$Z_2 = \frac{Z_3}{K_1} \tag{6.103a}$$

$$Z_1 = \frac{Z_2}{K_2} \qquad (6.103b)$$

The physical length and the width of the transmission lines in the trisection SIR are found using microstrip line equations. The symmetrical SIR illustrated in Figure 6.62 has

$$\theta_1 = 2\theta, \quad \theta_2 = \theta, \quad \theta_3 = \theta \qquad (6.104)$$

The physical length for each section in the SIR can be found from

$$l_n = \frac{\lambda_n \theta_n}{2\pi}, \quad n = 1,2,3 \qquad (6.105)$$

The width of the sections in SIR is obtained from Equations 6.92 through 6.94. The performance of BPFs with SIRs can be improved by using the configuration given in Figure 6.63. The BPF in Figure 6.63 provides triple band filter characteristics with the coupling scheme shown in Figure 6.64. In Figure 6.63, the coupled lines' equivalent circuit is represented by two single transmission lines of electrical length θ, and characteristic impedance Z_o and admittance inverter parameter J as shown in Figure 6.65. Inverter parameter J is an important design parameter because it is directly proportional to the coupling strength of the coupled lines.

This parameter is found using the network synthesis form of the equivalent circuit and is given by

$$J_{01} = Y_0 \sqrt{\frac{2k\theta_0}{g_0 g_1}} \qquad (6.106a)$$

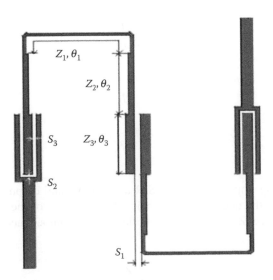

FIGURE 6.63 Triple band bandpass filter using SIRs.

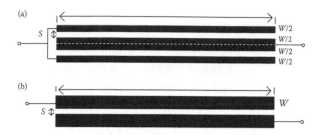

FIGURE 6.64   Coupling schemes. (a) Improved coupling scheme. (b) Conventional coupling scheme.

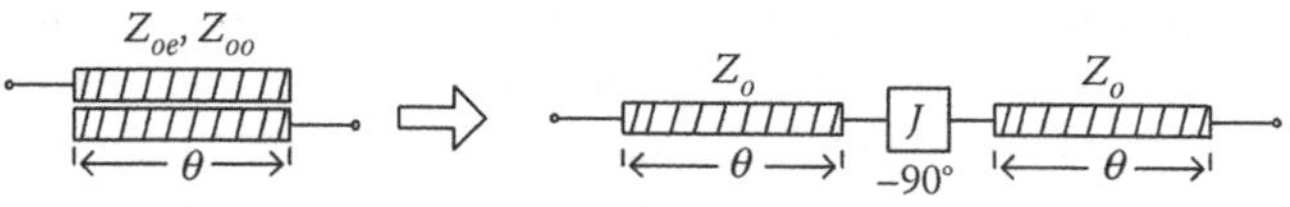

FIGURE 6.65   Equivalent circuit of parallel coupled lines.

$$J_{j,j+1} = Y_0 \frac{2k\theta_0}{\sqrt{g_j g_{j+1}}} \qquad (6.106b)$$

$$J_{n,n+1} = Y_0 \sqrt{\frac{2k\theta_0}{g_n g_{n+1}}} \qquad (6.106c)$$

As the ratio J/Y between the coupled lines increases, the coupling strength also increases.

Design Example: Stepped Impedance Resonator Bandpass Filters

Design a triple band BPF with SIRs using the improved coupling scheme. The center frequencies for each band are defined to be 1, 2.4, and 3.6 GHz. Use RO 4003 as a substrate with 32 mil thickness and 3.38 dielectric constant. The insertion loss in the passbands is required to be −3 dB or better. The return loss in the first and the second bands is desired to be −20 dB or lower. The third band stopband attenuation is specified to be −30 dB or lower. The ripple in the passband should not exceed 0.1 dB.

SOLUTION

The calculated design parameters for the given resonant frequency in each band using Equations 6.98 through 6.105 are given in Table 6.5.

TABLE 6.5

Design parameters for triple band microstrip bandpass filter

Given parameters	f_o (GHz)	f_{s1} (GHz)	f_{s2} (GHz)	D (mil)	ε_r	Z_3 (Ω)
	1	2.4	3.6	32	3.38	50
Calculated parameters	K_1	K_2	Z_1	Z_2		
	0.714	0.75	93.3	69.99		
Calculated physical dimensions	l_1 (in)	l_2 (in)	l_3 (in)	w_1 (mil)	w_2 (mil)	w_3 (mil)
	1.044	0.512	0.502	22.57	41.48	74.1

The layout of the filter using the dimensions in Table 6.5 is shown in Figure 6.66.

The filter that is shown in Figure 6.66 is simulated with the method of moment-based planar electromagnetic simulator, Sonnet, and built using the dimensions illustrated in Figure 6.66 with Roger 4003. The final version of the filter that is constructed is illustrated in Figure 6.67. 50 Ω SubMiniature version A (SMA) connectors are used for termination at the ports of the filter. The filter performance is measured with the network analyzer, Agilent 8753ES. The simulation and measurement results showing the overall performance of the filter up to 4 GHz for insertion loss |S21| and return loss |S11| are illustrated in Figure 6.68. The results for the insertion loss are close in the passband and deviates slightly from each other in the stopband. Figures 6.69

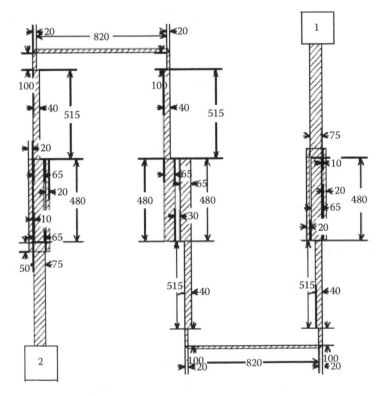

FIGURE 6.66   Layout of the triple band bandpass filter.

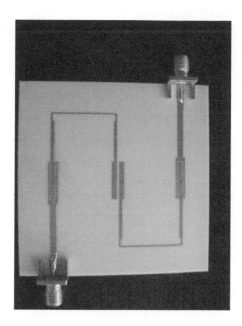

FIGURE 6.67 The constructed triple band trisection bandpass filter using SIRs.

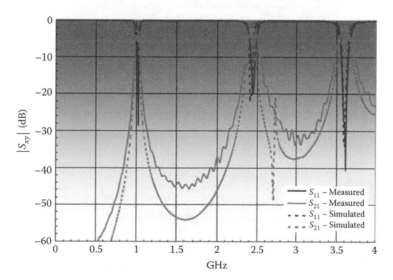

FIGURE 6.68 Measured and simulation results for insertion loss and return loss up to 4 GHz.

through 6.71 give a closer look to the filter performance in the first, second, and third frequency bands. The most deviation between the simulated and measured results in the passband is observed in the second band at 2.4 GHz. The simulated and measured results for the insertion loss are tabulated in Table 6.6. On the basis of the simulation and measurement results, the insertion loss specification is met except in the second

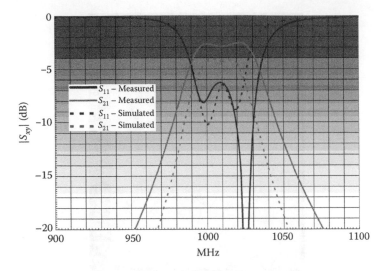

FIGURE 6.69 Filter performance in the first frequency band.

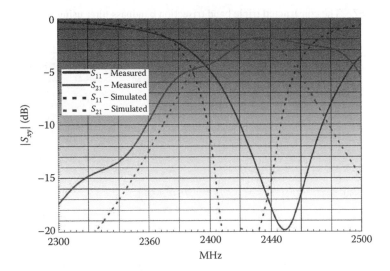

FIGURE 6.70 Filter performance in the second frequency band.

band where it is slightly lower. The simulation and measured results for the return loss are also found to be very close. The return loss specification is met in all the frequency bands as illustrated in Figures 6.68 through 6.71.

The effect of coupling between each resonator on the filter performance is studied using planar electromagnetic simulator for three different cases in each frequency band. These cases represent different coupling distances between SIRs and are designated by g. The coupling distance, g, is set to be 10, 30, and 60 mil. The simulation

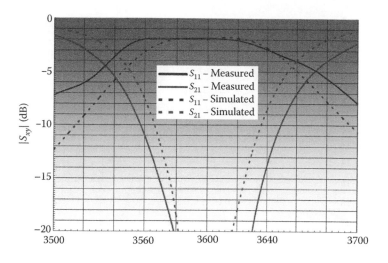

FIGURE 6.71   Filter performance in the third frequency band.

TABLE 6.6
Tabulation of Measurement and Simulation Results

Frequency (GHz)	Simulation (dB)	Measurement (dB)
1	4	3
2.4	3	4.5
3.6	2	2

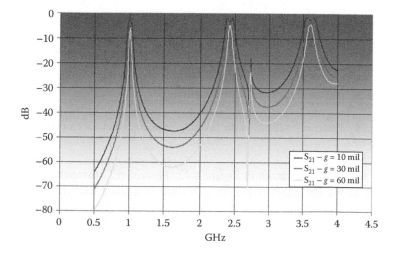

FIGURE 6.72   Coupling effect between SIRs on insertion loss up to 4 GHz.

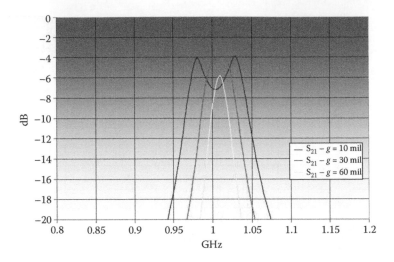

FIGURE 6.73 Effect of coupling in the first frequency band for the trisection triple band bandpass filter.

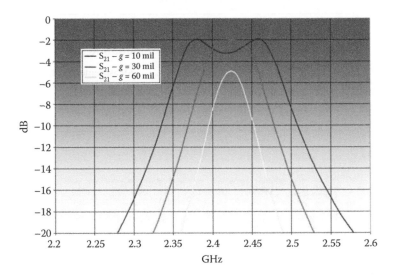

FIGURE 6.74 Effect of coupling in the second frequency band for trisection triple band bandpass filter.

results up to 4 GHz are shown in Figure 6.72. Figures 6.73 through 6.75 present a closer look to see the effect of coupling in each frequency band for the insertion loss. It has been observed that as the coupling gap between SIRs is decreased, wider bandwidth is obtained for each frequency band in the passband. However, although the wider bandwidth is obtained for the minimum coupling spacing, it results in ripples in the passband. As a consequence, the insertion loss is decreased more than 2 dB at the center frequency of the first frequency band. As the coupling spacing

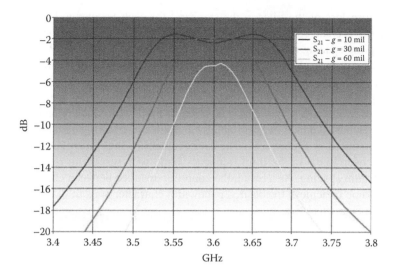

FIGURE 6.75 Effect of coupling in the third frequency band for trisection triple band band-pass filter.

increases, the bandwidth gets narrower and the insertion loss also decreases. The coupling spacing 30 mil between the coupled lines gives optimum filter performance in all frequency bands in terms of bandwidth and insertion loss level. This shows that although the minimum coupling spacing between the coupling segments gives the highest coupling level, it does not necessarily give the best filter performance since the insertion loss characteristics gets worse unlike the improvements in the bandwidth.

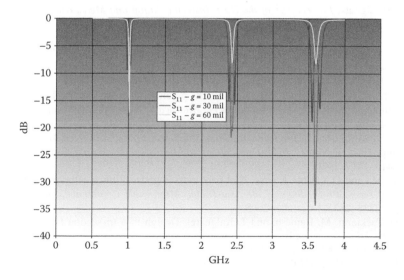

FIGURE 6.76 Coupling effect between SIRs for return loss up to 4 GHz.

The return loss of the filter in the stopband for each frequency band when the coupling gap is changed from 10 to 60 mil shows different characteristics. The minimum coupling spacing results in resonances in all frequency bands. When the coupling spacing is increased to 60 mil, the return loss is improved in the first frequency band, but gets worse in the second and third frequency bands. So, in the second and third frequency bands, the return loss of the filter is improved as the gap distance decreases whereas this phenomenon reverses in the first frequency band. The best overall filter performance is again obtained when the coupling spacing is 30 mil. This is illustrated in Figure 6.76. The improvement in the return loss with 30 mil spacing is over 15 dB in the third frequency band with respect to return loss level with minimum spacing. As a result, the coupling distance between each SIR can be used as a knob to achieve the desired bandwidth characteristics in the passband and attenuation characteristics.

REFERENCES

1. G.L. Matthaei, L. Young, and E.M.T. Jones. *Microwave Filters, Impedance Matching Networks, and Coupling Structures*, Artech House, Dedham, MA, 1980.
2. E.O. Hammerstard. Equations for microstrip circuit design, *Proceedings of the European Microwave Conference,* Hamburg, Germany, pp. 268–272, 1975.
3. H. Zhang and K.J. Chen. A tri-section stepped-impedance resonator for cross-coupled bandpass filters, *IEEE Microwave and Wireless Components Letters*, 15(6), 401–403, 2005.
4. I.C. Hunter, L. Billonet, B. Jarry, and P. Guillon. Microwave filters—Applications and technology, *IEEE Transactions on Microwave Theory and Techniques*, 50(3), 794–805, 2002.
5. R. Saal and E. Ulbrich. On the design of filters by synthesis, *IRE Transactions, CT-5*, 284–327, 1958.
6. L. Vanbeylen and J. Schoukens. Comparison of filter design methods to generate analytic signals, *Proceedings of the IEEE Instrumentation and Measurement Technology Conference (IMTC)*, Sorrento, Italy, pp. 24–27, April 2006.
7. V. Crnojevic-Bengin and D. Budimir. Design of thick film microstrip low pass filters, *IEEE International Conference on Telecommunications in Modern Satellite, Cable and Broadcasting Service*, Nis, Serbia and Montenegro, TELSIKS, pp. 360–362, October 2003.
8. A.B. Dalby. Interdigital microstrip circuit parameters using empirical formulas and simplified model, *IEEE Transactions on Microwave Theory and Techniques*, MTT-27, 744–752, 1979.
9. M. Makimoto and S. Yamishita. Bandpass filters using parallel coupled stripline stepped impedance resonators, *IEEE Transactions on Microwave Theory and Techniques*, 28(2), 1413–1417, 1980.
10. W. Ma and Q.X. Chu. A novel dual-band step-impedance filter with tunable transmission zeros, *Proceedings of the Asia-Pacific Microwave Conference*, Suzhou, China, pp. 2833–2835, December 2005.
11. D. Packiaraj, M. Ramesh, and T. Kalghatgi. Design of a tri-section folded SIR filter, *IEEE Microwave and Wireless Component Letters*, 16(5), 317–319, 2006.
12. E.G. Cristal and S. Frankel. Design of hairpin-line and hybrid haripin-parallel-coupled-line filters, *IEEE MTT-S Digest*, 1, 12–13, 1971.

7 MF-UHF RFID System Design Techniques

7.1 INTRODUCTION

Radio frequency identification devices (RFIDs) have been used widely in industrial and medical applications. Specifically, RFIDs have been applied in asset tracking and security systems by several companies extensively. RFID can store and retrieve data using RFID tags/transponders remotely. The conventional RFID system includes tags, reader, local software/infrastructure, and back-end system. RFID tag is the identification device in the system and contains at least a microchip attached to an antenna that sends data to the RFID reader. The antenna emits radio signals to activate the tag and to read and write to it. The signal strength depends upon the power output. When the RFID tag passes the electromagnetic zone, it detects the reader's activation signal. The reader decodes the data encoded in the tag's integrated circuit (IC) and passes it to host the computer for processing as shown in Figure 7.1. RFID tags can be active, passive, or semipassive tags. Active RFID tags are powered and as a result, require continuous power for operation. These types of tags have better communication distance. Semipassive tags use power for internal tag circuitry; however, they establish communication using the RF signal energy. Passive tags completely operate and communicate based on the energy of RF signal and hence very limited communication distance.

The frequency of operation and typical communication distances for RFID tags are given in Table 7.1. Passive tags are very popular for several applications due to the capacity of operation without a power source; however, their limited communication distance is the main challenge with RFID systems. There is extensive research being conducted to increase the communication distance on the existing systems with passive tags. One way to increase the RFID communication distance is to improve the performance of the antenna used in the system. RFIDs are commonly used with sensors to acquire and measure physical variables such as temperature, pressure, tensile

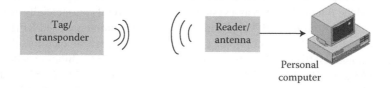

FIGURE 7.1 Basic block diagram of RFID system.

309

TABLE 7.1

Frequency Band, Applications, and Communication Distances for RFID Tags

RFID Frequency Band	Frequency Band	Typical Communication Distance	Common Application
			Animal tracking
			Access control
125–134.2 and 140–148.5 kHz	Low frequency	Up to ~1/2 m	Product authentication
			Smart cards
13.553–13.567 MHz	High frequency	Up to ~1 m	Shelve item tracking
			Airline baggage
			Maintenance
			Pallet tracking
			Carton tracking
858–930 MHz			Electronic toll collection
902–928 MHz, North America	Ultrahigh frequency	Up to 10 m	Parking lot access
			Electronic toll collection
2.45 GHz/5.8 GHz	Microwave	Up to 2 m	Airline baggage

strength, and so on during the operation of electronic and mechanical devices and materials. Sensors are vital components to capture and measure the critical parameters during the operation of any device. This technique is also used in biomedical applications. In this chapter, the basic RFID system design will be discussed and the RFID tag antenna design using microstrip technology will be detailed. The enhancement of the performance of microstrip antennas with electromagnetic structures and sensor-enabled RFIDs will also be discussed.

7.2 BASIC RFID SYSTEM DESIGN

A passive RFID tag consists of two critical components: microchip and antenna. The microchip is an integrated circuit that is used for storing and processing information, modulating and demodulating the RF signal, and other required functions. The antenna is used to receive electromagnetic waves from the reader's signal to energize the tag and send and receive data. The antenna is attached to the microchip via the matching network and the filter to function as a system. There are several commercially available chips that can be used for RFID application. Two of the commonly used RFID chips are TI CC1101 and TI CC2533 [9]. TI CC1101 has good performance for ultrahigh frequency range (UHF) RFID applications whereas TI CC2533 exhibits excellent performance for 2.4 GHz IEEE 802.15.4 applications. The layout of TI CC2533 is shown in Figure 7.2. The antenna is placed between pins 25 and 26 of TI CC2533, which are referred as RF_P and RF_N on the chip layout in Figure 7.2. This chip exhibits an optimized differential impedance between pins 25 and 26 to the RF system that is

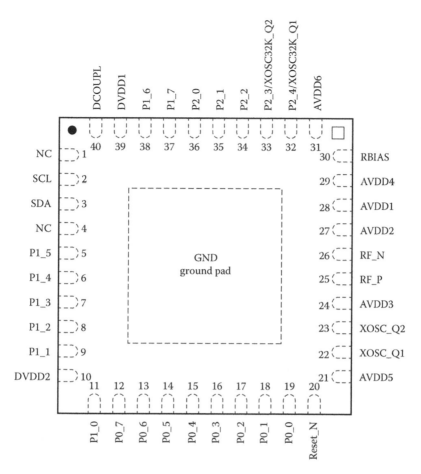

FIGURE 7.2 Layout of TI CC2533 chip.

interfaced at the frequency of interest. It is a common practice to give optimized differential impedance rather than individual RF port impedances on the data sheet of RFID chips. This has to be taken into account when interfacing an antenna and the associated matching network to these pins for accurate operation. Otherwise, mismatch will happen and the desired RF signal will not appear at the differential ports.

An example of an RFID system that is developed by TI using TI CC2533 chip is shown in Figure 7.3. In the design of the RFID system such as the one shown in Figure 7.3, the first task is to identify the optimized differential impedance of the chip using the data sheet. The data sheet of TI CC2533 lists the optimized differential impedance as

$$Z_{op} = 69 + j29 \; [\Omega] \tag{7.1}$$

Impedance matching is accomplished using discrete components and the antenna impedance is assumed to be 50 Ω. The RFID circuit shown in Figure 7.3 is simulated

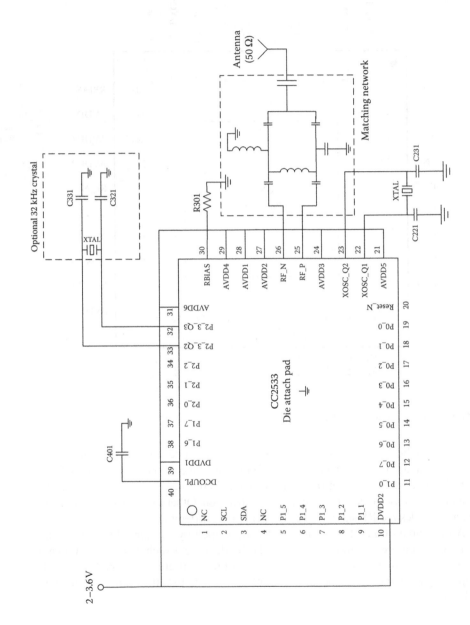

FIGURE 7.3 RFID system using TI CC2533 chip.

to obtain the characteristics of TI CC2533 at the differential ports and at the point where the antenna is interfaced. The RFID system in Figure 7.3 is accomplishing unbalanced to balanced impedance transformation. In an ideal system, the impedance at the port RF_N would be equal to the impedance at the port RF_P, which requires each port impedance to be equal to

$$Z_{op} = 34.5 + j14.5 \ [\Omega] \tag{7.2}$$

As a result, the phase difference between the two ports becomes 180°

$$\theta_{diff} = \theta_{RF\_P} - \theta_{RF\_N} = 180° \tag{7.3}$$

The total signal at the differential port then becomes balanced and in phase with the input RF signal received from the antenna. This results in a matched condition and maximum signal strength at the differential ports. However, the impedances at the differential ports of TI CC2533 are not symmetrical. They exhibit different impedances to the antenna and the matching should be implemented accordingly. The characteristic of the circuit shown in Figure 7.3 can be obtained using the frequency domain circuit simulator such as Ansoft Designer based on S-parameters. The simulated equivalent circuit for the RFID system in Figure 7.3 is shown in Figure 7.4. The simulation will be used to determine the input impedance of the antenna for a matched condition.

The differential port impedance between pins 25 and 26 is used to represent the RFID chip. It is important to note that ideal components are used in the simulation and the effect of printed circuit board (PCB) is not taken into account. The

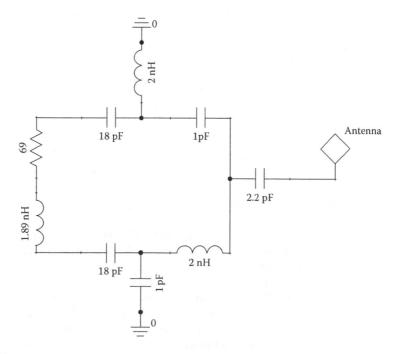

FIGURE 7.4   RFID equivalent circuit simulation.

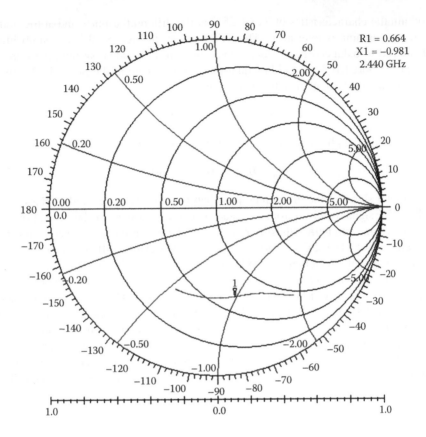

FIGURE 7.5   Input impedance of the antenna without PCB parasitics.

simulated input impedance at the operational frequency, $f = 2.44$ GHz, is found to be $33.2 - j49.05\ \Omega$ without the effect of PCB parasitics shown in Figure 7.5. This impedance is used to determine the PCB parasitics to match the required antenna impedance, which is $50\ \Omega$. The effect of PCB can be taken into account by an additional lumped LC matching network. The new circuit takes into account the PCB parasitics as illustrated in Figure 7.6. The antenna input impedance is now matched and the antenna input impedance versus the frequency is given in Figure 7.7. The return loss is shown in Figure 7.8 and is equal to -41.46 dB. The high return loss is an indication of a good RFID system design.

On the basis of the impedance-matching technique, the impedance at each port can also be determined. This has been applied and the impedance at each differential port for the operational frequency of interest as shown in Figure 7.9 is obtained. The return loss for the circuit shown in Figure 7.9 is found to be -40.48 dB at $f = 2.44$ GHz.

Instead of applying impedance matching with discrete components as explained in detail for the RFID system, balun can also be used. This helps in the reduction of the number of parts used in the design. One example circuit given by TI using balun for the UHF RFID system is given in Figure 7.10. The parts count is tremendously reduced as shown in Figure 7.10.

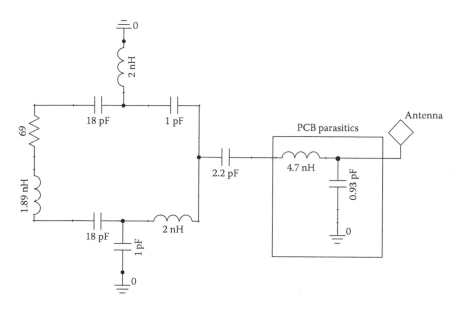

FIGURE 7.6 RFID equivalent circuit simulation.

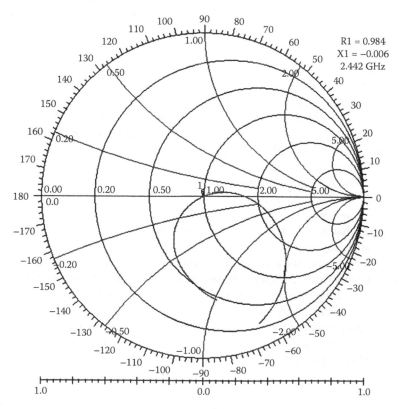

FIGURE 7.7 Input impedance of antenna with PCB parasitics.

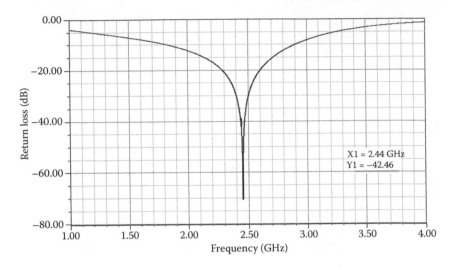

FIGURE 7.8   Return loss of RFID system with PCB parasitic effects.

7.3   RFID MICROSTRIP PATCH ANTENNA DESIGN

Microstrip patch antennas are widely used RFID antennas because of their various advantages such as low profile, light weight, easy fabrication, and conformability to mounting hosts in addition to size reduction and bandwidth enhancement [1–4]. The reading distance of microstrip patch antenna depends on several factors, including tag antenna gain, radar cross section, operating frequency, and antenna design parameters such as physical characteristics of the substrate used. The UHF for RFID passive antennas give an optimum reading distance for the operating conditions in comparison to the existing RFID systems as shown in Table 7.1. The pros and cons of the use of microstrip patch antennas are given in Table 7.2.

A patch antenna is designed based on the required resonant frequency (f_r), size, and bandwidth. The size and bandwidth requirements determine the dielectric constant (ε_r) and height (h) of the substrate. Increasing the height of the substrate increases the bandwidth, but it also increases the size of the antenna and could increase the propagation of surface waves, which causes performance degradation. Meanwhile, increasing the dielectric constant decreases the size of the antenna but narrows the bandwidth. Therefore, the substrate's dielectric constant and height must be selected carefully depending on the application. The geometry of the microstrip patch antenna is given in Figure 7.11.

The design procedure for the rectangular patch antenna begins with dielectric substrate selection. Once a dielectric substrate is selected, the width (W) and length (l) of the radiating patch can be calculated using the relation

$$W = \frac{c}{2f_r}\left(\frac{2}{\varepsilon_r + 1}\right)^{1/2} \tag{7.4}$$

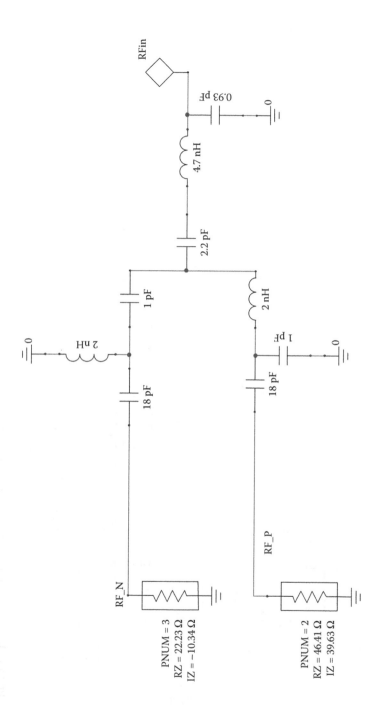

FIGURE 7.9 Impedance determination at each differential port.

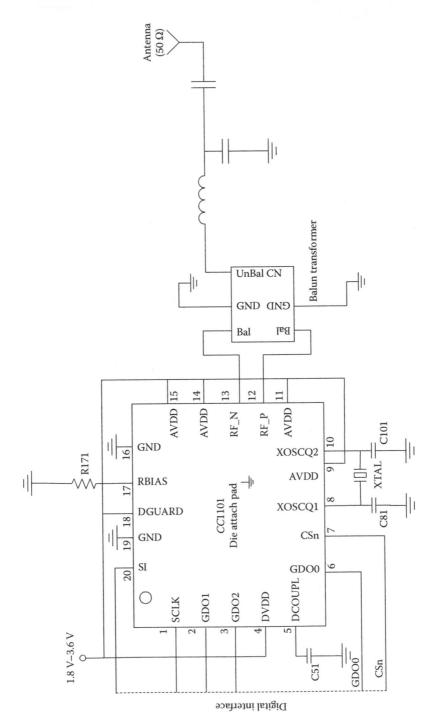

FIGURE 7.10 RFID system using TI CC1101 chip with balun.

TABLE 7.2

Pros and Cons of Microstip Patch Antennas

Pros	Cons
Light weight	Narrow bandwidth
Low volume	Not the greatest gain
Low profile	Incapable of handling high power
Conformal to multiple surfaces	Extraneous radiation from junctions and feeds
Low fabrication cost	Excitation of surface waves

(a)   (b)

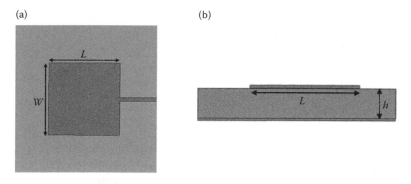

FIGURE 7.11 Geometry of microstrip patch antenna. (a) Top view and (b) side view.

After a width is selected, the length of the patch can be determined by the following relations:

$$\varepsilon_{eff} = \frac{\varepsilon_r + 1}{2} + \frac{\varepsilon_r - 1}{2}\left(1 + \frac{12h}{W}\right)^{-1/2} \quad (7.5)$$

$$\Delta l = 0.412h\frac{(\varepsilon_{eff} + 0.3)(W/h + 0.264)}{(\varepsilon_{eff} - 0.258)(W/h + 0.8)} \quad (7.6)$$

The length of the patch is then found from

$$l = \frac{c}{2f_r\sqrt{\varepsilon_{eff}}} - 2\Delta l \quad (7.7)$$

When the square patch antenna is designed, the width and the length of the patch are found from

$$l = \frac{c}{2f_r\sqrt{\varepsilon_{eff}}} - 2*0.412h\frac{(\varepsilon_{eff} + 0.3)(l/h + 0.264)}{(\varepsilon_{eff} - 0.258)(l/h + 0.8)} \quad (7.8)$$

Equation 7.8 can be reorganized as follows:

$$0 = F_1(l) - F_2(l) - F_3(l) \tag{7.9}$$

where

$$F_1(l) = \frac{c}{2f_r}\left(\frac{\varepsilon_r + 1}{2} + \frac{\varepsilon_r - 1}{2\sqrt{1 + 12(h/l)}}\right)^{-1/2} \tag{7.10}$$

$$F_2(l) = 0.824h\frac{\left[\left((\varepsilon_r + 1/2) + (\varepsilon_r - 1/2\sqrt{1 + 12(h/l)})\right) + 0.3\right](l/h + 0.264)}{\left[\left(\varepsilon_r + 1/2 + (\varepsilon_r - 1/2\sqrt{1 + 12(h/l)})\right) - 0.258\right](l/h + 0.8)} \tag{7.11}$$

and

$$F_3(l) = l \tag{7.12}$$

A numerical technique needs to be used to obtain the values of width and length of the patch. One of the optimization technique is to find the width and length of the patch by finding the minimum of the absolute value of Equation 7.9. The minimum value of a function can be found by using an algorithm such as the Nelder–Mead algorithm. To obtain desirable return loss at the resonant frequency, a microstrip patch antenna must be matched to the transmission line feeding it. There are two ways to match the patch antenna to the transmission line. The first way to match the patch antenna to the transmission line is to adjust the location of the feed, as shown in Figure 7.12. The second way is to use a quarter-wave transformer as a matching element, as shown in Figure 7.13. The input impedance of the patch antenna at the resonant frequency must be calculated before matching is implemented. The transmission line model and cavity model can be applied to calculate the input impedance at the edge of the patch antenna. Once the edge input impedance is calculated, the different matching techniques can be employed. In the transmission line model, the patch antenna is viewed as two radiating slots separated by a low-impedance transmission line that is approximately $\lambda/2$ in length. To obtain the resonant input impedance, one may start by finding the conductance of one of these slots. The conductance for a slot of finite width may be approximated as

$$G_1 = \frac{W}{120\lambda_0}\left[1 - \frac{1}{24}(k_0h)^2\right]\frac{h}{\lambda_0} < \frac{1}{10} \tag{7.13}$$

The input impedance at resonant frequency using the transmission line model is then calculated from

$$Z_{in} = \frac{1}{Y_{in}} = \frac{1}{2G_1} \tag{7.14}$$

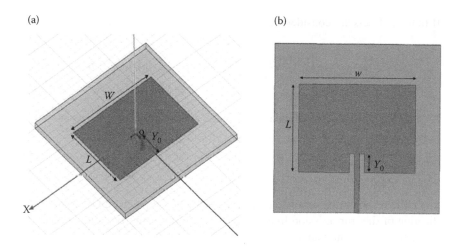

FIGURE 7.12 (a) Probe-fed patch antenna. (b) Inset-fed patch antenna.

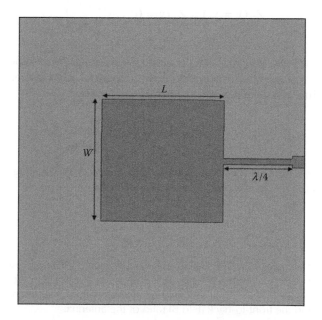

FIGURE 7.13 Patch antenna fed with a quarter-wave transformer.

In the cavity model, the patch antenna is viewed as "an array of two radiating narrow apertures (slots)" separated by a distance of approximately $\lambda/2$. The conductance of a slot is given by

$$G_1 = \frac{1}{120\pi^2}\left[-2 + \cos(k_0 W) + k_0 W \, \varepsilon \, S_i(k_0 W) + \frac{\sin(k_0 W)}{k_0 W}\right] \quad (7.15)$$

If mutual effects are considered, the mutual conductance is found from

$$G_{12} = \frac{1}{120\pi^2} \int_0^\pi \left[\frac{\sin((k_0 W/2)\cos\theta)}{\cos\theta} \right]^2 J_0(k_0 l \sin\theta) \sin^3\theta \, d\theta \qquad (7.16)$$

As a result, the input impedance of the patch antenna using the cavity model is

$$Z_{in} = \frac{1}{2(G_1 \pm G_{12})} \qquad (7.17)$$

Since the resonant input impedance at the edge of the patch can be found by the application of the transmission line or cavity modeling using Equation 7.14 or 7.17, the matching of the antenna to a feeding transmission line can be done by adjusting the location of the feed using

$$Z_{in}(y = y_0) = Z_{in}(y = 0) \, \cos^2\left(\frac{\pi y_0}{L}\right) \qquad (7.18)$$

Matching the patch antenna with the quarter-wave transformer is done by implementing a microstrip transmission line with a length of $\lambda/4$ between the feeding transmission line and the patch antenna, which is given by

$$Z_{\lambda/4} = \sqrt{Z_o Z_L} \qquad (7.19)$$

In Equation 7.19, Z_o is the characteristic impedance of the feeding transmission line, and Z_L is the resonant input impedance at the edge of the patch.

Example 1

Design and simulate a UHF RFID microstrip patch antenna resonating at $f = 915$ MHz with minimum 5 dB gain at the resonance using high dielectric permittivity substrate with $\varepsilon_r = 10.2$. Obtain the radiation pattern, the return loss versus frequency, and the front-to-back (F to B) ratio of the antenna.

SOLUTION

Using the method outlined and application of the relations, the microstrip patch antenna for UHF is designed. The dimensions of the antenna are given in Table 7.3.

The structure is simulated using HFSS. The simulated patch antenna is shown in Figure 7.14.

The simulation results showing the radiation patterns versus frequency are given in Figures 7.15 and 7.16. Figures 7.15 and 7.16 give the radiation patterns in

TABLE 7.3

Dimensions of Patch Antenna

Substrate		Patch		Feed Location	
Height	10 mm	Length	46.65 mm	y_0	14.925 mm
Dielectric constant	10.2	Width	46.65 mm		
Length	130 mm				
Width	130 mm				

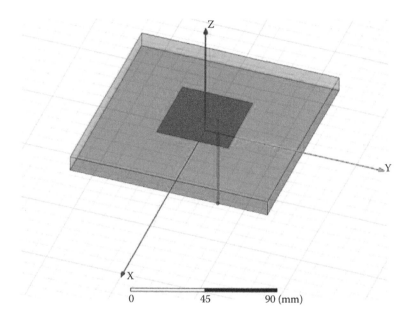

FIGURE 7.14 Simulated UHF RFID microstrip patch antenna.

rectangular and polar coordinate systems on XY and YZ planes. The return loss of the antenna is given in Figure 7.17.

The maximum gain at the resonant frequency is found to be 5.68 dB. The return loss and F to B ratio of the microstrip patch antenna are −48.17 and 9.55 dB. The simulation results of the antenna at the resonant frequency are tabulated in Table 7.4.

7.4 RFID MICROSTRIP PATCH ANTENNA DESIGN WITH EBG STRUCTURES

One way to improve the radiation characteristics of a patch antenna is the use of electromagnetic band gap (EBG) structures [5–8]. EBG structures are periodic structures that reduce the propagation of surface waves. The reduction of surface waves increases antenna gain, minimizes the back lobe, which increases directivity, and

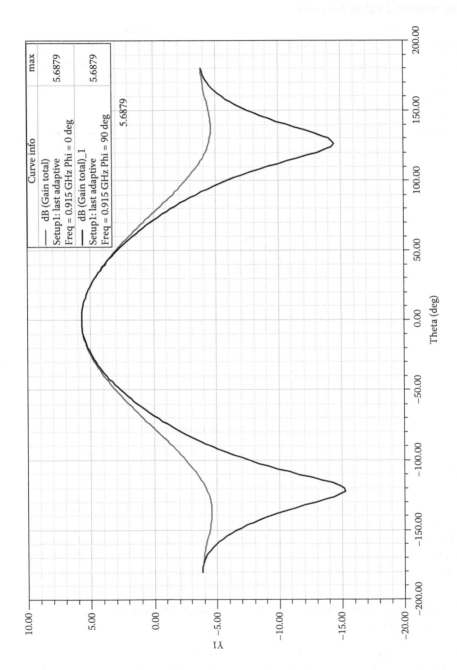

FIGURE 7.15 2D radiation pattern of the antenna in the rectangular coordinate system.

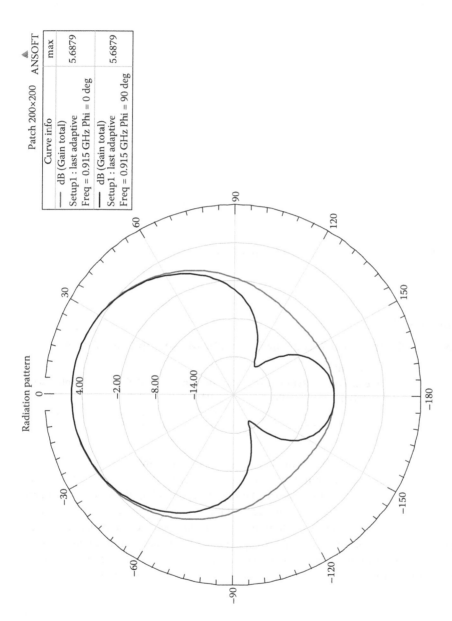

FIGURE 7.16 2D radiation pattern of the antenna in the polar coordinate system.

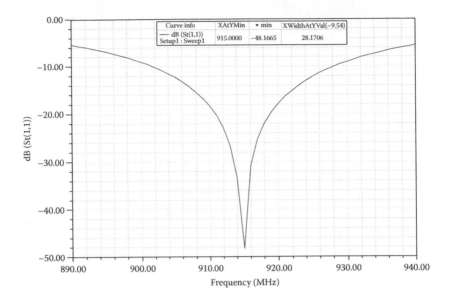

FIGURE 7.17   Simulated return loss of UHF RFID microstrip patch antenna.

TABLE 7.4
Summary of Simulation Results for UHF RFID Microstrip Patch Antenna

	Resonant Frequency (MHz)	Return Loss (dB)	Bandwidth (MHz)	Peak Gain (dB)	Front-to-Back (dB)
Patch antenna	915	−48.17	28.17	5.6879	9.5497

increases the bandwidth. An example of microstrip patch antenna with mushroom-like EBG structure is shown in Figure 7.18. Figure 7.19 shows a cross section of the mushroom-like EBG structure. When the operating wavelength is large compared to the periodicity of the mushroom-like EBG structure, the EBG structure can be approximated by an effective medium model with lumped LC elements. The small gaps between the patches generate a capacitance, and the current along adjacent patches produces an inductance as shown in Figure 7.20. The impedance of the LC model of the EBG structure shown in Figure 7.18 can be found using

$$Z = \frac{j\omega_r L}{1 - \omega_r^2 LC} \tag{7.20}$$

As the frequency comes closer toward the results of Equation 7.21, the impedance increases toward infinity, creating a frequency band gap:

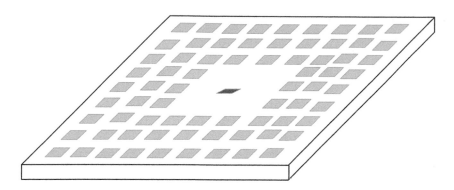

FIGURE 7.18 Microstrip patch antenna with mushroom-like EBG structure.

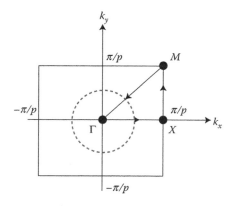

FIGURE 7.19 Brillouin zone.

$$f_r = \frac{1}{2\pi\sqrt{LC}} \tag{7.21}$$

The capacitance and inductance can be found through the following equations:

$$C = \frac{W\varepsilon_0(1 + \varepsilon_r)}{\pi}\cosh^{-1}\left(\frac{W + g}{g}\right) \tag{7.22}$$

$$L = \mu_0\mu_r h \tag{7.23}$$

When Equations 7.22 and 7.23 are substituted into Equation 7.21, the resonant frequency can be written as

$$f_r = \frac{c}{2\pi}\left(\frac{W\mu_r h(1 + \varepsilon_r)}{\pi}\cosh^{-1}\left(\frac{W + g}{g}\right)\right)^{-1/2} \tag{7.24}$$

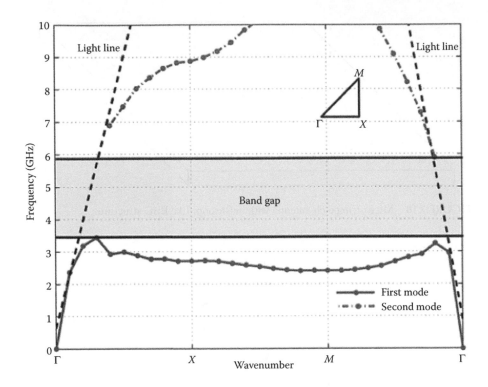

FIGURE 7.20   Dispersion diagram for mushroom-like EBG structure.

Equation 7.24 can be reorganized by gathering known values on one side and unknown values on the other side of the equation as given by

$$\frac{c^2}{4\pi f_r^2 \mu_r h(1 + \varepsilon_r)} = W\cosh^{-1}\left(\frac{W + g}{g}\right) \tag{7.25}$$

The width (W) of the square mushroom top can be selected and the gap (g) can be found through

$$g = \frac{W}{\cosh^{-1}(c^2/4\pi f_r \mu_r h(1 + \varepsilon_r)W) - 1} \tag{7.26}$$

The structure can also be designed based on its periodic length

$$a = W + g \tag{7.27}$$

If a periodic length is found, the length of the EBG structure's gap can be obtained from Equations 7.25 and 7.27 as

$$0 = (a - g)\cosh^{-1}\left(\frac{a}{g}\right) - \frac{c^2}{4\pi f_r^2 \mu_r h(1 + \varepsilon_r)} \tag{7.28}$$

After the length of the gap is determined, the width of the EBG structure is computed using Equation 7.27. The relationship between the bandwidth of the frequency band gap and the lumped LC elements for the EBG structure is given by

$$BW \propto \frac{L}{C} \tag{7.29}$$

Using the inductance and capacitance that are calculated with Equations 7.22 and 7.23, an approximation for the bandwidth of the frequency band gap can be obtained as

$$BW = \frac{f_r}{120\pi}\sqrt{\frac{L}{C}} \tag{7.30}$$

The important criteria in the implementation of EBG structures is to have the band gap centered at the given resonant frequency. The frequency of the surface wave band gap can be determined using mainly three methods: dispersion method, reflection phase method, and direct transmission method. The three methods are compared using three-dimensional (3D) electromagnetic simulator HFSS, and theoretical and simulated results for the surface wave band frequency are tabulated in Table 7.5. The dispersion diagram method gives the closest surface wave band frequency to the resonant frequency as shown in Table 7.5.

The dispersion diagram technique is applied with the identification of Brillouin zone points shown in Figures 7.19 and 7.20 using

$$\Gamma : k_x = 0, \quad k_y = 0 \tag{7.31}$$

$$X : k_x = \frac{2\pi}{W + g}, k_y = 0 \tag{7.32}$$

TABLE 7.5

Comparison of Surface Wave Band Frequencies with Different Methods

Method	Band Gap Frequency (MHz)	Center Frequency (MHz)	Bandwidth (MHz)
Theoretical	827.5–1002.5	915	175
Dispersion diagram	843–987	915	144
Reflection phase	772.3–942.7	857.5	170.4
Direct transmission	915–1255	1085	340

TABLE 7.6

Dimensions of Patch Antenna with EBG Structures

Substrate		Patch		Feed Location		EBG	
Height	10 mm	Length	46.5 mm	y_0	16.4	Width	23.6 mm
Dielectric constant	10.2	Width	46.5 mm			Gap	2.01 mm
Length	130 mm					Radius of Vias	1 mm
Width	130 mm					Distance from the patch	16.17 mm
						Number of units	16
						Number of rows	1

and

$$M: k_x = \frac{2\pi}{W + g}, \quad k_y = \frac{2\pi}{W + g} \tag{7.33}$$

Example 2

Design and simulate a UHF RFID microstrip patch antenna in the previous example with EBG structures and compare radiation pattern, return loss versus frequency, and F to B ratio of the antenna by keeping all the design constraints the same.

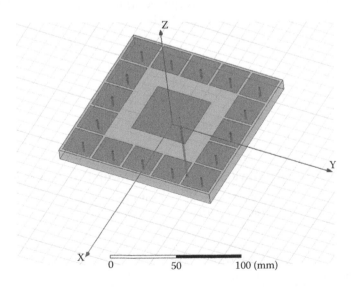

FIGURE 7.21   Simulated UHF RFID microstrip patch antenna with EBG structure.

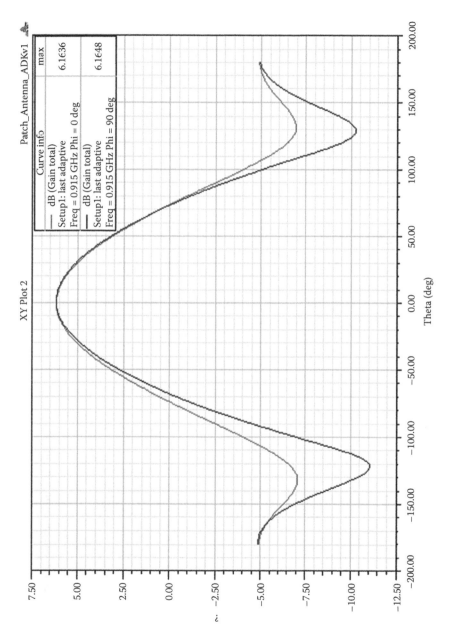

FIGURE 7.22 Patch antenna gain with EBG structure on XY and YZ planes in rectangular coordinate system.

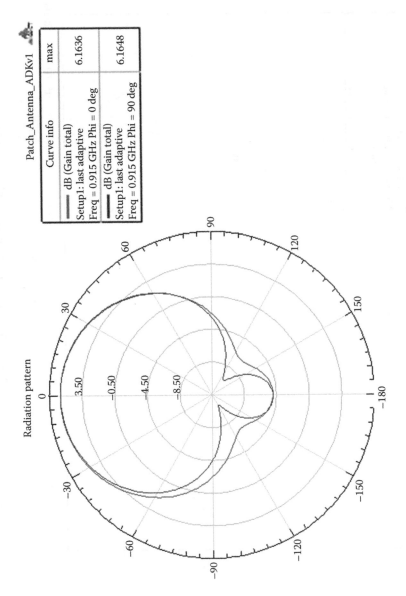

FIGURE 7.23 Patch antenna gain with EBG structure on XY and YZ planes in polar coordinate system.

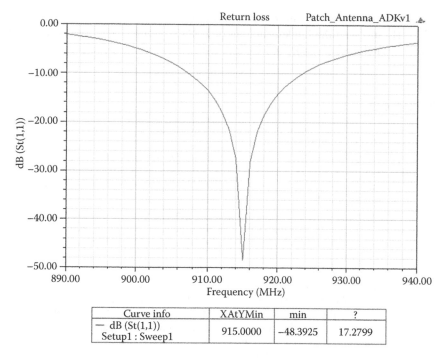

Curve info	XAtYMin	min	?
— dB (St(1,1)) Setup1 : Sweep1	915.0000	−48.3925	17.2799

FIGURE 7.24 Simulated return loss of patch antenna with EBG structure.

TABLE 7.7
Comparison of Results for Microstrip Antenna with and without EBG Structures

	Resonant Frequency (MHz)	Return Loss (dB)	Peak Gain (dB)	Front-to-Back (dB)
Patch antenna	915	−48.17	5.6879	9.5497
Patch with EBG	915	−48.39	6.1648	11.0656

SOLUTION

The physical dimensions of the patch antenna with EBG structures are given in Table 7.6. The antenna with EBG structures is simulated again with HFSS and the simulation results for radiation pattern, return loss, and F to B ratio are obtained. The simulated structure is given in Figure 7.21.

The radiation patterns for microstrip patch antenna with EBG structures in rectangular and polar coordinate systems on XY and YZ planes are shown in Figures 7.22 and 7.23. The return loss of the antenna is given in Figure 7.24. It is shown that there is a 0.5 dB improvement in the maximum gain and a 1.5 dB improvement in the F to B ratio with the implementation of EBG structures. The results for RFID microstrip patch antenna with and without EBG structures are compared and tabulated in Table 7.7.

REFERENCES

1. G. Breed, The fundamentals of patch antenna design and performance, *High Frequency Electronics*, 8(3), 48–52, 2009.
2. C.A. Balanis, *Antenna Theory: Analysis and Design*, 3rd ed., John Wiley & Sons, New York, 2005.
3. D.G. Fang, *Antenna Theory and Microstrip Antennas*, CRC Press Taylor & Francis Group, Boca Raton, FL, 2009.
4. T.T. Nguyen, D. Kim, S. Kim, and J. Jang, Design of a wideband mushroom-like electromagnetic bandgap structure with magneto-dielectric substrate, *6th International Conference on Information Technology and Applications*, Hanoi, Vietnam, pp. 130–135, November 2009.
5. B. Gao and M.M.F. Yuen, Passive UHF RFID with ferrite electromagnetic band gap (EBG) material for metal objects tracking, *Electronic Components and Technology Conference*, Lake Beuna Vista, FL, pp. 1990–1994, May 27–30, 2008.
6. M.N. Tan, T.A. Rahman, S.K.A. Rahim, M.T. Ali, and M.F. Jamlos, Antenna array enhancement using mushroom-like electromagnetic band gap (EBG), *Antennas and Propagation (EuCAP), 2010 Proceedings of the Fourth European Conference*, Barcelona, Spain, pp. 1–5, April 2010.
7. S.M. Moghadasi, Waveguide model for reflection phase characterization of periodic EBG surfaces, *Asia Pacific Microwave Conference*, Bangkok, Thailand, pp. 1–4, December 11–14, 2007.
8. F. Yang and Y.R. Samii. *Electromagnetic Band Gap Structures in Antenna Engineering*. Cambridge University Press, Cambridge, UK, 2009.
9. Texas Instruments, http://www.ti.com/.

Index